THE MULTISENSC

T0227717

Human Factors in Road and Rail Transport

Series Editors

Dr Lisa Dorn
*Director of the Driving Research Group, Department of Human Factors,
Cranfield University*

Dr Gerald Matthews
Professor of Psychology at the University of Cincinnati

Dr Ian Glendon
*Associate Professor of Psychology at Griffith University, Queensland,
and is president of the Division of Traffic and Transportation Psychology
of the International Association of Applied Psychology*

Today's society must confront major land transport problems. The human and financial costs of vehicle accidents are increasing, with road traffic accidents predicted to become the third largest cause of death and injury across the world by 2020. Several social trends pose threats to safety, including increasing car ownership and traffic congestion, the increased complexity of the human-vehicle interface, the ageing of populations in the developed world, and a possible influx of young vehicle operators in the developing world.

Ashgate's 'Human Factors in Road and Rail Transport' series aims to make a timely contribution to these issues by focusing on the driver as a contributing causal agent in road and rail accidents. The series seeks to reflect the increasing demand for safe, efficient and economical land-based transport by reporting on the state-of-the-art science that may be applied to reduce vehicle collisions, improve the usability of vehicles and enhance the operator's wellbeing and satisfaction. It will do so by disseminating new theoretical and empirical research from specialists in the behavioural and allied disciplines, including traffic psychology, human factors and ergonomics.

The series captures topics such as driver behaviour, driver training, in-vehicle technology, driver health and driver assessment. Specially commissioned works from internationally recognised experts in the field will provide authoritative accounts of the leading approaches to this significant real-world problem.

The Multisensory Driver
Implications for Ergonomic Car Interface Design

CRISTY HO

Daiwa Scholar, The Daiwa Anglo-Japanese Foundation, UK

&

CHARLES SPENCE

*Professor of Experimental Psychology, Head of the Crossmodal Research
Laboratory, Oxford University, UK*

CRC Press
Taylor & Francis Group
Boca Raton London New York

CRC Press is an imprint of the
Taylor & Francis Group, an **informa** business

CRC Press
Taylor & Francis Group
6000 Broken Sound Parkway NW, Suite 300
Boca Raton, FL 33487-2742

First issued in paperback 2017

© 2008 by Cristy Ho and Charles Spence
CRC Press is an imprint of Taylor & Francis Group, an Informa business

No claim to original U.S. Government works
Version Date: 20160226

ISBN 13: 978-1-138-07584-9 (pbk)
ISBN 13: 978-0-7546-7068-1 (hbk)

Visit the Taylor & Francis Web site at
http://www.taylorandfrancis.com

and the CRC Press Web site at
http://www.crcpress.com

Contents

List of Figures

List of Tables

Acknowledgements

The research reported in this book was supported in part by an Overseas Research Students Award and a Clarendon Fund bursary, from Oxford University, as well as by a Senior Scholarship from Somerville College, Oxford University to Cristy Ho. We would like to thank our collaborators Prof. Hong Tan at Purdue University, Dr. Nick Reed at TRL (Transport Research Laboratory), and Dr. Valerio Santangelo at University of Rome 'La Sapienza'. We would also like to express our gratitude to the TRL for hosting and setting up the two simulator trials reported in Experiments 5.3 and 7.3.

Chapter 1

Introduction

The attentional limitations associated with driving

Humans are inherently limited capacity creatures; that is, they are able to process only a small amount of the sensory information that is typically available at any given time (see Broadbent, 1958; Driver, 2001; Simons and Chabris, 1999; Spence and Driver, 1997a). Although it is unclear what the exact nature of this limited capacity is (see Posner and Boies, 1971; Schumacher et al., 2001), the inability of humans to simultaneously process multiple sources of sensory information places a number of important constraints on their attentional processing of stimuli both in the laboratory and in a number of real-life settings (e.g., see McCarley et al., 2004; O'Regan, Rensink, and Clark, 1999; Spence and Read, 2003; Velichkovsky et al., 2002).

Over the years, attention has been defined in a number of different ways. It has, for example, been defined as the '*ability to concentrate perceptual experience on a selected portion of the available sensory information to achieve a clear and vivid impression of the environment*' (Motter, 2001, p. 41). One very important element of any definition of attention, however, is its selectivity (see Driver, 2001, for a review). Among all of the various different dimensions along which the selective processing of information may take place, the spatial distribution of attention represents an area that is of great interest to many researchers, from both a theoretical and an applied standpoint (see Posner, 1978; Spence, 2001; Spence and Driver, 1994; Spence and Read, 2003).

The limited capacity of spatial attention to process sensory information in humans raises important constraints on the design and utilization of, for instance, vehicular information systems (e.g., Brown, Tickner, and Simmonds, 1969; Burke, Gilson, and Jagacinski, 1980; Chan and Chan, 2006; Mather, 2004; Spence and Driver, 1997a). The act of driving represents a highly complex skill requiring the sustained monitoring of integrated perceptual and cognitive inputs (Hills, 1980). The ability of drivers to attend selectively and their limited ability to divide their attention amongst all of the competing sensory inputs have a number of important consequences for driver performance. This, in turn, links inevitably to the topic of vehicular accidents. For instance, a driver may fail to detect the sudden braking of the vehicle in front if distracted by a passenger's conversation (or by the conversation with someone at the other end of a mobile phone; e.g., Horrey and Wickens, 2006; Spence and Read, 2003; see Chapter 2), resulting in a collision with the vehicle in front (see Sagberg, 2001; Strayer and Drews, 2004). In fact, one recent research report has shown that the presence of two or more car passengers is associated with a two-fold increase in the likelihood of a driver having an accident as compared to people who drive unaccompanied (see McEvoy, Stevenson, and Woodward, 2007b). In addition, these attentional limitations on driver performance are currently being exacerbated by the ever-increasing availability of complex in-vehicle technologies (Ashley, 2001; Lee, Hoffman, and Hayes, 2004; Wang, Knipling, and Goodman, 1996; though see also Cnossen, Meijman, and Rothengatter, 2004), such as satellite navigation systems (e.g., Burnett and Joyner, 1997; Dingus, Hulse, Mollenhauer, Fleischman, McGehee, and Manakkal, 1997; Fairclough, Ashby, and Parks, 1993), mobile phones (e.g., Jamson, Westerman, Hockey, & Carstens, 2004; Patten, Kircher, Ostlund, and Nilsson, 2004; Spence and Read, 2003; Strayer, Drews, and Johnston, 2003), email (e.g., Harbluk and Lalande, 2005; Lee, Caven, Haake, and Brown, 2001) and ever more elaborate sound systems (e.g., Jordan and Johnson, 1993). Somewhat surprisingly, this proliferation of in-vehicle interfaces has taken place in the face of extensive research highlighting the visual informational overload suffered by many drivers (e.g., Bruce, Boehm-Davis, and Mahach, 2000; Dewar, 1988; Dukic, Hanson, and Falkmer, 2006; Hills, 1980) and the widely reported claim in the literature that at least 90 per cent of the information used by drivers is visual (e.g., Booher, 1978; Bryan, 1957; Spence & Ho, forthcoming; though see also Sivak, 1996).

Given the many competing demands on a driver's limited cognitive resources, it should come as little surprise that driver inattention, including drowsiness, distraction and 'improper outlook', has been identified as one of the leading causes of vehicular accidents, estimated to account for anywhere between 26 per cent (Wang et al., 1996) and 56 per cent (Treat et al., 1977) of all road traffic accidents (see also Ashley, 2001; Gibson and Crooks, 1938; Klauer, Dingus, Neale, Sudweeks, and Ramsey, 2006; McEvoy et al., 2007; Sussman, Bishop, Madnick, and Walter, 1985). Fortunately, however, the last few years have also seen the development of a variety of new technologies, such as sophisticated advanced collision avoidance systems, designed to monitor the traffic environment automatically, and to provide additional information to drivers in situations with a safety implication. In fact, the development of these new technologies, known collectively as intelligent transport systems (ITS; Noy, 1997), means that more information than ever before can now potentially be delivered to drivers in a bid to enhance their situation awareness and ultimately improve road safety.

One of the most common types of car accident, estimated to account for around a quarter of all collisions, is the front-to-rear-end (FTRE) collision (McGehee, Brown, Lee, and Wilson, 2002; see also Evans, 1991). The research that has been published to date suggests that driver distraction represents a particularly common cause of this kind of accident (no matter whether the lead vehicle is stationary or moving; Wang et al., 1996; see also Rumar, 1990). Research by Strayer and Drews (2004) has also shown that mobile phone use tends to be one of the factors leading to front-to-rear-end collisions (see also Alm and Nilsson, 2001; Sagberg, 2001).

The potential benefits associated with improving the situation awareness of drivers to road dangers such as impending collisions are huge. To put this into some kind of perspective, it has been estimated that the introduction of a system that provided even a modest decrease in the latency of overall driver responses of, say, around 500 ms, would reduce front-to-rear-end collisions by as much as 60 per cent (Suetomi and Kido, 1997; see also 'The mobile phone report', 2002). Given that a number of different forward collision warning systems now exist, it has become increasingly important to determine the optimal means of assisting drivers to avoid such collisions (e.g., Graham, 1999). Current strategies differ in terms of their degree of intervention, varying from pro-active collision avoidance systems that can initiate automatic emergency braking responses, to collision warning systems that simply present warning signals to drivers in an attempt to get them to adjust their speed voluntarily instead (Hunter, Bundy, and Daniel, 1976; Janssen and Nilsson, 1993).

In fact, the technology now exists to enable 'intelligent' cars to detect dangerous road situations on the road ahead (e.g., such as adaptive radar cruise control systems and computerized safety systems that enable cars to communicate with each other prior to potential collisions), which in theory means that the cars of the future could potentially soon take control away from the driver and become autonomous (Knight, 2006). However, at present, car manufacturers and governmental organizations appear to prefer that the control of the car remains in the hands of the drivers (though see Smith, 2008). The primary reason for this is related to the legal implications and issues arising should an accident occur during automatic cruising, such as those relating to liability for the accident (see Hutton and Smith, 2005; Knight, 2006). Therefore, investigations into the design of optimal warning signals that can alert drivers to potential dangers are essential given the increasing availability of advanced driver assistance systems (see Yomiuri Shimbun, 2008).

A great deal of empirical effort has gone into studying how best to alert and warn inattentive drivers of impending road dangers. For example, a number of researchers have suggested that non-visual (e.g., auditory and/or tactile) warning signals could be used in interface design (e.g., Deatherage, 1972; Hirst and Graham, 1997; Horowitz and Dingus, 1992; Lee et al., 2004; Sorkin, 1987; Stokes, Wickens, and Kite, 1990; see Table 1.1).

Multisensory integration

It is our belief that one increasingly important way in which to design effective in-car interfaces is by understanding the limitations of the human brain's information processing system (see Spence and Ho, forthcoming). This area of research is known as 'neuroergonomics' (e.g., see Fafrowicz and Marek, 2007; Sarter & Sarter, 2003). Traditional theories of human information processing have typically considered individual sensory modalities, such as audition, vision and touch, as being independent. In particular, the most influential multiple resource theory (MRT) on human workload and performance originally proposed by Christopher Wickens in 1980 (see e.g., Wickens, 1980, 1984, 1992, 2002) more than a quarter of a century ago specifically postulated the existence of independent pools of attentional resources for the processing of visual and auditory information (Hancock, Oron-Gilad, and Szalma, 2007; see Figure 1.1). According to the multiple resources account, when people are engaged in concurrent tasks that consume their visual and auditory resources simultaneously, such as talking on the mobile phone while driving, there should be no dual-task cost due to the putative independence of the attentional resources concerned. That is, according to Wickens' theory, the conversation is processed

Table 1.1   Advantages of non-visual (auditory and/or tactile) warning signals for interface design

Purported advantages of non-visual over visual warning signals	Study
Responses to auditory and tactile signals typically more rapid than to visual signals.	Jordan (1972); Nelson, McCandlish, and Douglas (1990); Todd (1912)
Non-visual warning signals should not overload the driver's 'visual' system (cf. Sivak, 1996).	McKeown and Isherwood (2007); though see Spence and Driver (1997b); Spence and Read (2003)
Inherently more alerting.	Campbell et al. (1996); Gilmer (1961); Posner, Nissen, and Klein (1976)
Non-visual warnings do not depend for their effectiveness on the current direction of a driver's gaze. They are perceptible even when the drivers' eyes are closed, such as during blinking; or when the visual system is effectively 'turned off', such as during saccades that typically occur several times a second.	Bristow, Haynes, Sylvester, Frith, and Rees (2005); Hirst and Graham (1997); Stanton and Edworthy (1999)
Non-visual warning signals tend to be judged as less annoying than many other kinds of warning signal.	Lee et al. (2004); McGehee and Raby (2003)

by the auditory-verbal-vocal pathway, while the act of driving is processed by the visual-spatial-manual pathway. Therefore, these two processes should not, in theory, conflict with each other.

The evidence that has emerged from both behavioural and neuroimaging studies over the last decade or so has, however, argued against this traditional view (see Navon, 1984). Instead, the recently developing multisensory approach to human information processing has put forward the view that people typically integrate the multiple streams of sensory information coming from each of their senses (e.g., vision, audition, touch, olfaction and taste) in order to generate coherent multisensory perceptual representations of the external world (e.g., see the chapters in Calvert, Spence, and Stein, 2004; Spence and Driver, 2004). In fact, the multisensory integration of inputs from different sensory modalities appears to be the norm, not the exception. What is more, the research published to date has shown that multisensory integration takes place automatically under the majority of experimental conditions (see Navarra, Alsius, Soto-Faraco, and Spence, forthcoming, for a review). Linking back to the limited capacity of human information processing, interference or bottlenecks in attention could therefore occur at at least two different stages, specifically, at the modality-specific level and/or at the crossmodal level that is shared between different sensory modalities (as well as at the perceptual; see Lavie, 2005, and also at the response selection stages; see Levy, Pashler, and Boer, 2006; Pashler, 1991; Schumacher et al., 2001).

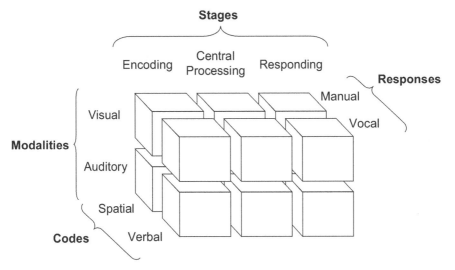

Figure 1.1   **Schematic diagram of the Multiple Resource Theory of human workload and performance proposed by Wickens (e.g., 1980, 2002)**

The extensive body of evidence that has emerged from recent laboratory-based research has provided support for the existence of robust crossmodal links in spatial attention between different sensory modalities (see Spence and Driver, 2004; Spence and Gallace, 2007). In particular, the available research now suggests that the efficiency of human multisensory information processing can be enhanced if the relevant information provided to the different senses is presented from approximately the same spatial location (see Driver and Spence, 2004) at approximately the same time (Spence and Squire, 2003). The research has also shown that it is normally harder to selectively attend to one sensory signal if a concurrent (but irrelevant) signal coming from another sensory modality is presented from approximately the same location (see Spence and Driver, 1997b; Spence, Ranson, and Driver, 2000b). This also implies that it may be harder to ignore a sensory signal in one sensory modality if it is presented at (or near) the current focus of a person's spatial attention in another sensory modality (see Spence et al., 2000b).

Crossmodal links in spatial attention have now been observed between all possible combinations of auditory, visual and tactile stimuli (Spence and Driver, 1997a, 2004). These links in spatial attention have been shown to influence both exogenous and endogenous attentional orienting (Klein, 2004; Klein and Shore, 2000; Posner, 1980; see Driver and Spence, 2004; Spence, McDonald, and Driver, 2004, for reviews). Exogenous orienting refers to the stimulus-driven (or bottom-up) shifting of a person's attention whereby the reflexive orienting of attention occurs as a result of external stimulation. By contrast, endogenous orienting refers to the voluntary shifting of a person's attention that is driven internally by top-down control. A variety of laboratory-based research has suggested that independent mechanisms may control these two kinds of attentional orienting (e.g., Berger, Henik, and Rafal, 2005; Hopfinger and West, 2006; Klein and Shore, 2000; Santangelo and Spence, forthcoming; Spence and Driver, 2004).

Current approaches to the design of auditory warning signals

Over the years, a number of different approaches to the design of effective auditory warning signals have been proposed. These include the use of spatially-localized auditory warning signals (e.g., Begault, 1993, 1994; Bliss and Acton, 2003; Bronkhorst, Veltman, and van Breda, 1996; Campbell et al., 1996; Humphrey, 1952; Lee, Gore, and Campbell, 1999), the use of multisensory warning signals (e.g., Hirst and Graham, 1997; Kenny, Anthony, Charissis, Darawish, and Keir, 2004; Mariani, 2001; Mowbray and Gebhard, 1961; Selcon, Taylor, and McKenna, 1995; Spence and Driver, 1999; see also Haas, 1995; Spence and Ho, forthcoming) and the use of synthetic warning signals (such as auditory earcons; e.g., Lucas, 1995; McKeown and Isherwood, 2007) that have been artificially engineered to deliver a certain degree of perceived urgency.

To date, these various different approaches have met with mixed success (see Lee et al., 1999; cf. Rodway, 2005). For example, researchers have found that people often find it difficult to localize auditory warning signals, especially when they are presented in enclosed spaces such as inside a car, hence often negating any benefit associated with the spatial attributes of the warning sound (see Bliss and Acton, 2003; Fitch, Kiefer, Hankey, and Kleiner, 2007). Meanwhile, other researchers have reported favourably on the potential use of directional sounds in confined spaces (e.g., Cabrera, Ferguson, and Laing, 2005; Catchpole, McKeown, and Withington, 2004).

Moreover, it often takes time for interface operators to learn the arbitrary association between a particular auditory earcon and the appropriate response, as the perceived urgency is transmitted by the physical characteristics of the warning signal itself (such as the rate of presentation, the fundamental frequency of the sound and/or its loudness, etc.; e.g., Edworthy, Loxley, and Dennis, 1991; Graham, 1999; Haas and Casali, 1995; Haas and Edworthy, 1996). It would appear, therefore, that unless auditory earcons can somehow be associated with intuitive responses on the part of the driver (Graham, 1999; Lucas, 1995), they should not be used in dangerous situations to which an interface operator (and, in particular, the driver of a road vehicle) may only rarely be exposed, since they may fail to produce the appropriate actions immediately (see also Guillaume, Pellieux, Chastres, and Drake, 2003).

Given these limitations in the use of traditional warning signals, a number of researchers have attempted to investigate whether auditory icons (i.e., sounds that imitate real-world events; Gaver, 1986) might provide more effective warning signals, the idea being that they should inherently convey the meaning of the events that they are meant to signify (e.g., Blattner, Sumikawa, and Greenberg, 1989; Gaver, 1989; Gaver, Smith, and O'Shea, 1991; Lazarus and Höge, 1986; McKeown, 2005). Over the years, the effectiveness of a variety of different 'urgent' auditory icons has been evaluated in terms of their ability to capture a person's attention, and perhaps more importantly, to elicit the appropriate behavioural responses from them (e.g., see Deatherage, 1972; Oyer and Hardick, 1963).

Auditory icons have the advantage over auditory earcons in that their meaning should be more immediately apparent to an interface operator and so people should need less time in order to learn the appropriate behavioural responses to such signals (e.g., Begault, 1994; Lucas, 1995). However, while the research that has been published to date has shown that people do indeed tend to respond more rapidly to auditory warning signals as the perceived urgency increases (e.g., Burt, Bartolome, Burdette, and Comstock, 1995; Haas and Casali, 1995), the use of urgent auditory icons is not without its own problems. For example, while the screeching car tyre and car horn sounds used in a study reported by Graham (1999) elicited faster responses by drivers than the more typical tonal alert or verbal warning signals, the presentation of these auditory icons also resulted in participants making more inappropriate responses than following the tonal or verbal alerts. It would therefore appear that highly urgent signals may sometimes elicit such rapid responses that interface operators can end up responding to the warning signal before they have had sufficient time in which to evaluate the situation properly in order to know what the most appropriate response would have been (see Bliss and Acton, 2003; Graham, 1999). Furthermore, the empirical research that has been conducted to date does not provide unequivocal support for the claim that auditory icons necessarily make for particularly effective urgent warning signals.

The use of urgent auditory alarm icons may be further limited by the fact that auditory icons that are perceived as conveying a high degree of urgency are also likely to be perceived as unpleasant (McKeown and Isherwood, 2007; see also Oyer and Hardick, 1963). For example, in one recent study, McKeown and Isherwood assessed the perceived unpleasantness of twenty different environmental sounds. They found a strong correlation between the perceived urgency of the sounds and how unpleasant people rated those sounds as being (see Figure 1.2). Therefore, while the approach of trying to develop auditory warning signals that elicit an intuitive response from drivers would, at first glance, seem like a good one, the fundamental problem appears to be that any sound that is perceived by a person as being urgent will most likely also be

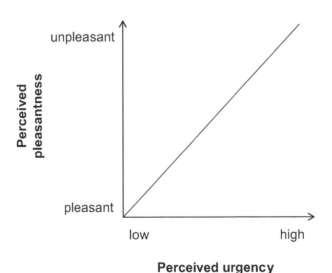

Perceived urgency

Figure 1.2 **Graph showing the linear relationship between the perceived degree of urgency of an auditory icon and its perceived degree of pleasantness**

judged as unpleasant, and thus will be unlikely to be accepted by the end user of an interface (see McKeown and Isherwood, 2007; cf. Lazarus and Höge, 1986).

One potentially interesting class of auditory warning signals is constituted by speech warnings. Verbal warning signals have the advantage that minimal training is required in order for an operator who understands the language to comprehend and act upon them efficiently (see Edworthy and Hellier, 2006; cf. Bruce, et al., 2000; Simpson, McCauley, Roland, Ruth, and Williges, 1987). As suggested by Edworthy and Hellier in their comprehensive review of complex auditory warnings, the intelligibility of speech warnings as perceived by listeners can be affected by the environment. Given that while driving, a person may be involved in other concurrent speech tasks such as conversing with the passenger or on the mobile phone, and/or listening to the radio (see Dibben and Willamson, 2007; Ho, Reed, and Spence, 2007b; Ramsey and Simmons, 1993), the utility of in-car speech warnings (particularly directional signals such as those given by navigation systems) may be somewhat limited (see Strayer and Johnston, 2001). Note also the limited utility of such warning signals for drivers whose first language happens to be different from the one in which the warning signal is presented.

With respect to the timing of warning signal presentation, McGehee et al. (2002) have assessed the effectiveness of presenting auditory signals at different times to warn distracted drivers of an impending collision with a stationary vehicle on the road ahead. They found that the presentation of an 'advance' auditory warning signal facilitated driver responses relative to either a 'late' warning signal (giving only 1 s, as opposed to 1.5 s, advance warning), or else to a no warning signal baseline condition in their driving simulator study. This facilitation of performance included shorter accelerator release times as well as fewer, and less severe, crashes. While results such as these highlight the potential benefits of presenting auditory warning signals to enhance the situation awareness of drivers, it is important to note that a number of potential limitations have also been identified with the presentation of 'early' warning signals. For instance, although the presentation of an early warning signal allows a driver more time in which to prepare and execute the appropriate behavioural response, research has shown that the more advance notice a warning signal provides to an interface operator of a potentially-dangerous upcoming event, the more likely it is that the signal will be classified as a false alarm (McGehee et al., 2002; Parasuraman, Hancock, and Olofinboba, 1997; Shinar, 1978). What is more, as soon as the false alarm rate for a warning signal becomes too high, interface operators may well begin to perceive it as being a nuisance and therefore start to ignore it. Thus, the danger is that interface operators may become desensitized to future warning signals (see Bliss and Acton, 2003; Breznitz, 1984). As Sorkin (1988) notes, some interface operators may even go so far as to try and disable alarm signals that they consider to be too distracting or aversive (see also King and Corso, 1993). It is therefore crucial that the collision warning signal is presented at the appropriate moment in time.

Collision avoidance

A two-second gap between a driven vehicle and a lead vehicle is recommended in the United Kingdom (Driving Standards Agency, 2004) in order to allow sufficient time for a car to stop in the case of an emergency. This minimum time-to-collision estimate has been broken down into 'thinking distance' (including perceptual processing) and 'braking distance' (i.e., the time required to program and execute a motor response). A review by Hills (1980) suggested that the 'thinking distance' required to brake while driving a car corresponds to a reaction time of 700 ms following the onset of a critical event (see also Campbell et al., 1996; Rockwell, 1988). This two-second rule has often been criticized as being inadequate in explaining the performance of the majority of drivers. Researchers have argued that the 85th percentile driver would require a standard of 3.2 s for event perception and braking reaction (see Dewar, 1988). Some researchers have even reported that 70 per cent of drivers involved in front-to-rear-end accidents failed to stop a car when the lead vehicle had stopped some 2–6 s prior to the crash incident (see Horowitz and Dingus, 1992).

In addition, in a meta-analysis of braking reaction time studies, Green (2000) reported that drivers had an average reaction time of 700 ms for a fully-expected event as compared to a reaction time of 1500 ms for an unexpected event. In his analysis, Green broke the total perception-brake reaction time into three major components, specifically, mental processing time (a combination of sensation, perception, and response selection and programming), movement time and device reaction time. While the reaction time of the device clearly constitutes a technical problem, some researchers have suggested that movement time (i.e., lifting the foot off the accelerator pedal in order to depress the brake pedal) can be improved on by training (see Levy et al., 2006). In any case, one key question in warning signal design is concerned with speeding up the mental processing of drivers (i.e., at the perceptual and decisional stages of human information processing).

Recent developments in tactile warning signal design

In addition to the traditional design of visual and auditory interfaces, the potential application of tactile warning signals and information displays in applied interface environments is currently receiving a great deal of both empirical and commercial interest (e.g., Fitch et al., 2007; Gallace, Tan, and Spence, 2007; Gilliland and Schlegel, 1994; Jones, Gray, Spence, and Tan, 2008; Jones and Sarter, forthcoming; Rupert, 2000b; Sklar and Sarter, 1999; Spence & Ho, submitted; van Erp and van Veen, 2004, 2006; Wood, 1998; Zlotnik, 1988). The communication of information by touch has been successfully demonstrated in a number of different application areas, such as in the aerospace domain to assist the spatial orientation of pilots and astronauts (e.g., Rochlis and Newman, 2000; Rupert, Guedry, and Reschke, 1994). For instance, somatosensory cues have been shown to help improve the situation awareness of astronauts, who otherwise rely mainly on visual cues, in confusing situations in altered sensory environments, such as under conditions of weightlessness (Rochlis and Newman, 2000).

Van Erp, Jansen, Dobbins, and van Veen (2004) reported two case studies demonstrating that directional and distance information can be communicated by using a vibrotactile torso display that allowed individuals to pilot a helicopter successfully, or to drive a high speed rigid inflatable boat in a waypoint navigation task. It should, however, be noted that despite these applied research efforts, no such systems are currently in place in aeroplanes. Instead, the automobile industry may represent a more pragmatic application domain (i.e., one less constrained by industrial standardization and regulatory control) where innovation may be easier to 'get through' the design process.

The rapid growth of interest in tactile interface design is supported by the older body of applied research on tactile sensation and perception dating back to the early work of Fenton in the 1960's. For example, in Fenton's (1966) study, a tactile control stick was used to present drivers with headway and relative velocity information via the tactile modality. Driving performance in a simulated (i.e., laboratory) car-following situation was facilitated when compared to performance using conventional automobile controls. The findings from a recent study by van Erp and van Veen (2004) also provide evidence to suggest a reduced subjective workload on the part of drivers when using tactile instead of visual navigation displays. Van Erp and van Veen studied the effectiveness of left/right vibrotactile cues presented to a driver's thigh via tactors embedded in the seat in terms of their ability to provide navigational messages to drivers under conditions of normal and high workload. However, although their participants responded more rapidly to bimodal navigational messages than to unimodal tactile messages, and vice versa in their other study (van Erp and van Veen, 2001), it should be noted that the comparison of their data in the bimodal and unimodal conditions unfortunately failed to reach statistical significance. Similarly, in another study of car drivers, Fitch et al. (2007) also failed to demonstrate that their participants found it any easier to respond to the location (from eight possible locations) that had been cued when audiotactile cues were compared to unimodal auditory (or tactile) cues. Overall, the findings of these two studies therefore fail to provide any strong support for the claim that redundant information presented simultaneously to different senses may prove useful in the design of multimodal, or multisensory, systems (see also Jackson and Selcon, 1997; Oviatt, 1999; Spence and Driver, 1999). Recent research by Santangelo and his colleagues (Santangelo and Spence, 2007; Santangelo, Ho, and Spence, 2008) has, on the other hand, shown that bimodal cues are significantly better than unimodal cues, at least when they are presented from the same spatial location (see Chapter 7).

It is worth noting that in-car interfaces with vibrating functions have already been implemented in several recently-released models of car (e.g., BMW 5 Series's vibrating steering wheel and Citroën C6's vibrating seat). However, limited published research has attempted to examine the potential beneficial or detrimental effects of such vibrotactile in-car systems on driving performance, and subsequently, on safety on the road. In particular, although the effects of tactile stimulation have been investigated on a wide array of different body parts, such as for stimuli presented to the torso (e.g., Ho and Spence, 2007; Jones et al., 2008; Linderman, Yanagida, Sibert, and Lavine, 2003), head (e.g., Gilliland and Schlegel, 1994), hands (e.g., Burke et al., 1980; Jagacinski, Miller, and Gilson, 1979; Vitense, Jacko, and Emery, 2003), wrists (e.g., Sklar and Sarter, 1999), buttocks (e.g., Lee et al., 2004; McGehee and Raby, 2003) and feet (e.g., Godthelp and Schumann, 1993; Janssen and Nilsson, 1993; Janssen and Thomas, 1997), no one has yet attempted to systematically examine the relationship between the body site stimulated and the nature of the driving task being (or to be) performed. Thus, it remains uncertain whether tactile stimulation, regardless of the body part being stimulated, will necessarily result in the same degree of improvement on a particular task. The alternative here is that optimal human performance in a particular task may be related to tactile stimulation on a specific body-site. So, for example, it might be the case that vibrating a driver's wrists represents the optimal means of getting them to perform an action with their hands. Note that research on stimulus-response compatibility effects suggests that there may indeed be important synergies between the stimulation of particular body parts and specific task requirements (e.g., see Fitts and Seeger, 1953; Kantowitz, Triggs, and Barnes, 1990; Proctor, Tan, Vu, Gray, and Spence, 2005). It is also worth mentioning at this point that it remains an open question as to whether tactile icons ('tactons'; for example, vibrations of various rhythm and/or roughness that can be reliably discriminated by people) exist (see Brewster and Brown, 2004; Brown, Brewster, and Purchase, 2005; see also Enriquez, MacLean, and Chita, 2006), as they might potentially provide a whole new range of possibilities for interface designers in terms of the artificial engineering of complex messages to convey information to users via the sense of touch (see Gallace et al., 2007, for a recent review).

Several studies have explicitly assessed the utility of vibrotactile cues in simulated driving scenarios. For example, Janssen and Thomas (1997) reported that increasing the counterforce (i.e., involving combined proprioceptive and tactile cuing) on the accelerator pedal had beneficial effects in a collision avoidance system (see also Janssen and Nilsson, 1993; Tijerina, Johnston, Parmer, Pham, Winterbottom, and Barickman, 2000). On the other hand, Schumann, Godthelp, Farber, and Wontorra (1993) contrasted various different kinds of tactile feedback with auditory warnings as a means of signalling to drivers whether to remain in lane when engaged in passing manoeuvres. They reported that drivers were more responsive to tactile or proprioceptive cues, such as the vibration of the steering wheel or force feedback cues resisting the driver's control input to change lanes, as compared to auditory warnings. Lee et al. (2004) have also compared the effectiveness of auditory and haptic collision warnings presented in either a graded or single-staged manner on drivers' braking performance. Their results showed that haptic warnings were judged as being less annoying, and that graded haptic alerts presented via a vibrating seat seemed to be particularly effective in providing a greater margin of driver safety.

Critical factors for effective collision warning signal design

It is important to note that it will not be sufficient for future in-car warning signals simply to be optimized for their ability to capture a driver's spatial attention (Spence, 2001). Instead, they must also be optimized for (i.e., be compatible with) the most appropriate driver response in any given situation. Typically, psychologists and other research scientists tend to divide behavioural phenomena into their perceptual, decisional, and response-related sub-components (e.g., see Kantowitz et al., 1990; Proctor et al., 2005; Spence, 2001). The future design of both unisensory and multisensory warning signals (which will be delineated in the chapters that follow) should therefore be optimized to elicit the most effective response by combining an understanding of the different brain pathways involved in information processing and the particular combinations of multisensory signals that are the most appropriate for specific classes of actions (e.g., for braking, accelerating, mirror checking, turning, etc.). This means that designers have to combine the available knowledge regarding each of the sub-processes involved in multisensory attention, event perception, response selection and response execution.

Taken together, the recent evidence that has emerged from both behavioural and neuroscience studies suggests that we may now be in a position where our understanding of the multisensory nature of human information processing is such that we can start to apply the insights gained from recent studies of crossmodal links in spatial attention to the design of applied multisensory interfaces (e.g., Spence and Driver, 1997a, 1999; Spence and Ho, 2008, forthcoming). In particular, one of the central aims of this book is to investigate the role of spatial attention in the domain of developing more ergonomic unisensory and multisensory warning signals for car drivers (i.e., to try and develop signals that elicit more efficient and effective responses from car drivers). Given that it has been argued that driving is predominantly a visual task (cf. Sivak, 1996), it is important to examine whether the use of non-visual warning signals can be used to effectively orient a driver's visual attention to the appropriate location. The neuroscience approach to multisensory attentional capture by warning signals has been made possible by the resolution in the experimental psychology literature that there exist robust crossmodal links in spatial attention between audition, vision and touch (see Spence and Driver, 2004; Spence and Gallace, 2007). However, it is also important to acknowledge the potential performance trade-off that may be associated with the utilization of additional sensory modalities of information to car drivers. In particular, laboratory-based research has shown that there is a cost associated with having to divide one's attention between multiple different sensory modalities at the same time (Spence and Driver, 1997; Spence, Nicholls, and Driver, 2001).

The chapters that follow in this book investigate the stage(s) in information processing at which the facilitatory effects elicited by such spatial multisensory warning signals occur (see Figure 1.3). For instance, it is particularly important to examine the relative improvements in information processing at the perceptual, decisional and/or motor response levels, in order to understand the underlying causes of any performance enhancement that is found.

The subsequent chapters are structured as follows: Chapters 2 and 3 provide reviews of the literature regarding the potential limitations in human information processing that constrain the performance of drivers who are distracted by trying to perform a secondary task. In particular, we focus on the consequences of talking on a mobile phone while driving (Chapter 2) and on listening to the radio (Chapter 3). On the basis of this review of the literature on distracted (and/or drowsy) drivers, we are able to highlight some of the key features that any attention-capturing warning signal needs to achieve if it is to effectively capture a driver's attention. In Chapter 4, we review the existing literature on auditory warning signals and report on the potential facilitation in responsiveness to collision situations with the presentation of spatial and/or meaningful auditory cues. Chapter 5 reviews the recent development

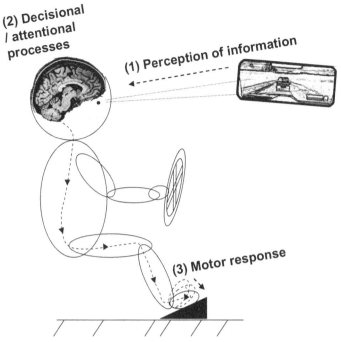

Figure 1.3 Schematic diagram showing the basic stages in information processing at which facilitatory effects on performance elicited by the presentation of spatial multisensory warning signals may occur

of vibrotactile in-car interfaces and discusses their purported 'intuitive' property. In Chapter 6, we provide an in-depth investigation into the relative contributions of perceptual enhancement and decisional facilitation to the effects discussed in the preceding experimental chapters. Chapter 7 further examines the spatial relationship between combined audiotactile (multisensory) cues and highlights the important role that multisensory integration can play in facilitating the disengagement of a person's spatial attention from a concurrent highly perceptually-demanding task. Finally, in Chapter 8, we highlight the most important findings to have emerged from the warning signal experiments that are described in this book. We then go on to discuss some of the possible directions for multisensory interface research in the coming years, including evaluating the potential for olfactory cues in modulating the behaviour and/or mental state of drivers.

Chapter 2

Driven to Distraction

Introduction

The evidence highlighted in the previous chapter clearly shows that there are many different forms of distraction that can lead to driver inattention, and hence possibly to vehicular accidents (see also Peters and Peters, 2001). However, the one that has been studied most thoroughly to date (and the one which has generated the most debate; see Eby, Vivoda, and St. Louis, 2006) concerns the possible adverse effects of talking (and, in particular, of using a mobile phone) while driving. It has been estimated that more than 75 per cent of the population of the U.K. own or use a mobile phone (see 'The mobile phone report', 2002). What's more, according to one survey conducted in the U.S. in 2004, 5 per cent of drivers (rising to 8 per cent of drivers between 16 and 24 years) will be talking on their handheld phone at any given moment during daylight hours (Glassbrenner, 2005; see also Johal, Napier, Britt-Compton, and Marshall, 2005; 'Phones and automobiles can be a dangerous mix', 1999). Eby et al. have estimated that this figure is set to rise to around 8.6 per cent for the general population by 2010. Hahn, Tetlock, and Burnett (2000, p. 48) have estimated that as much as 60 per cent of all mobile phone use in the U.S. occurs while drivers are at the wheel (though note that the figure may be somewhat lower in the U.K.; see 'The mobile phone report', 2002; see also Goodman et al., 1999). Meanwhile, more than 90 per cent of the 35,000 Canadian drivers who were questioned in another study reported having used their mobile phone while driving (see Laberge-Nadeau, Maag, Bellavance, Lapierre, Desjardins, Messier, and Saïdi, 2003).

A very large number of studies have now demonstrated that using a mobile phone while at the wheel can result in a significant impairment in driving performance (see Goodman et al., 1999; Horrey and Wickens, 2006, McCartt, Hellinga, and Bratiman, 2006, for reviews) and decision making (Alm and Nilsson, 1995; McKnight and McKnight, 1993). Therefore, in this chapter, we will review the laboratory, simulator, and on-road driving literature in order to try and understand the nature of the information processing limitations that constrain the performance of the typical driver when simultaneously trying to perform an auditory secondary task. We will then use the insights derived from this research to help determine exactly how the unisensory and multisensory warning signals that are to be evaluated in the chapters that follow should be designed in order to ensure that a driver's eyes, and more importantly (as we will see later), their visual attention, are focused on the relevant stimuli/events on the road around them.

Dual-task costs associated with using a phone while driving?

Redelmeier and Tibshirani (1997a) conducted one of the most widely-cited epidemiological studies to have assessed the risks associated with talking while driving. They examined the mobile phone records of 699 drivers in the Toronto area who had a mobile phone and who had also been involved in a car accident. The researchers assessed the number of mobile phone calls made by the drivers on the day of the crash and in the preceding week using a case-crossover design.[1] By using the pair-matched analytic approach, Redelmeier and Tibshirani were able to identify any increased risk should the drivers have made more phone calls immediately prior to the collision than would have been expected solely on the basis of chance.

The results of Redelmeier and Tibshirani's (1997a) study revealed that car drivers actually have a fourfold increased risk of having an accident if they use a mobile phone while driving. The results also suggested that the increase in risk was similar for both men and women (see also McKnight and McKnight, 1993; though see Violanti, 1997; Wetherell, 1981). More recently, McEvoy, Stevenson, McCartt, Woodward, Haworth, Palamara, and Cercarelli (2005) reported very similar results from a case-crossover study conducted in Perth, Western Australia of 456 drivers who had been involved in serious road accidents resulting in hospitalization (see also Laberge-Nadeau et al., 2003; Violanti, 1997, 1998; Violanti and Marshall, 1996; Young, 2001; though see Sullman and Baas, 2004). To put this figure into some kind of perspective, this increase in risk is equivalent to (if not greater than) the increase in accident risk seen when drivers exceed the drink-driving limit (see Vinson, Mabe, Leonard, Alexander, Becker, Boyer, and Moll, 1995; though it should be noted that these two factors may affect driving performance in somewhat different ways; Strayer and Drews, 2007; Strayer, Drews, and Couch, 2006; see also Redelmeier and Tibshirani, 1997b; 'The mobile phone report', 2002). In fact, the potential legal repercussions for drivers of admitting that they had been talking on the mobile phone when an accident took place means that these estimates concerning the risks associated with using a mobile phone while driving may even represent something of an underestimate (see Haigney and Westerman, 2001; James, 1991; McEvoy et al., 2005, on this point). Maclure and Mittleman (1997, p. 501) estimated (on the basis of Redelmeier and Tibshirani's data) that the financial cost to society from allowing the use of mobile phones while driving would be around $2–4 billion in the U.S. in the year 2000 alone.

One advantage of the case-crossover approach utilized by Redelmeier and Tibshirani (1997a) and others is that it allows one to rule out a number of alternative accounts of any increased accident risk that is observed, such as, for example, the possibility that anxious drivers simply have more crashes than other less-anxious drivers, and that anxious drivers are simply also more likely to have a phone in their cars. This approach also allows one to rule out the possibility that drivers who use

---

1   Note that this approach has the advantage that each person acts as their own control, thus ruling out the confounding influence of such factors as driver age, experience, sex, etc., on the results.

a mobile phone while driving may simply tend to take more risks than other drivers (Wilson, Fang, and Wiggins, 2003; see also Evans and Wasielewski, 1982). However, while the case-crossover approach can control for those factors that do not change over a short period of time, it cannot rule out the possibility that sudden events (such as, for example, a driver just having received some very bad news) might not have resulted in both a temporary increase in the number of phone calls that they might want to make, and to a worsening of their driving performance. The problem here then is one of pinpointing exactly those factors that govern the observed correlation between accidents and mobile phone use (see also Quinlan, 1997). Well-controlled laboratory studies have therefore helped researchers to get a better handle on what is actually going on and it is to these studies that we shall now turn (though as we will see later, such studies also suffer from their own problems; in particular related to the reduced ecological validity of laboratory, as compared to on-road, research; though see Haigney and Westerman, 2001; Nilsson, 1993).

Intuitively, one might think that the principal problem associated with using a mobile phone while driving is related to manual factors ('The mobile phone report', 2002): namely, to difficulties associated with trying to hold both the phone and the steering wheel at the same time (though see Mazzae, Ranney, Watson, and Wightman, 2004). Interestingly, only 27 per cent of drivers in one study thought that using a hands-free phone while driving was particularly distracting). Indeed, much of the legislation specifically designed to try and reduce mobile phone use in cars in the U.K. (and in North America; see Strayer and Drews, 2007) over the last few years has focused explicitly on increasing the penalties that are handed down to drivers who are caught using a handheld phone while at the wheel. In fact, while handheld phones are now illegal in many countries, hands-free phone use is still currently legal when driving in the U.K. and in much of North America (see McCartt et al., 2006; Sundeen, 2006). One might well ask whether such a position is really justified (see also Johal et al., 2005; McCartt, Braver, and Geary, 2003; McCartt and Geary, 2004).

Handheld versus hands-free phone use

The majority of the empirical research that has looked at this question has actually shown that using a hands-free mobile phone is no safer than using a handheld phone (e.g., Consiglio, Driscoll, Witte, and Berg, 2003; Horrey and Wickens, 2006; Lamble, Kauranen, Laasko, and Summala, 1999; McEvoy et al., 2005; Patten, Kircher, Östlund, and Nilsson, 2004; Redelmeier and Tibshirani, 1997a; Strayer and Johnston, 2001; though see also 'The mobile phone report', 2002). For example, Redelmeier and Tibshirani observed no significant difference between the accident rates of handheld and hands-free phone users in their epidemiological study, leading them to suggest that the dual-task deficit does not stem from a loss of manual dexterity, but is instead cognitive or attentional in nature.[2] This claim,

---

2   Note that Maclure and Mittleman (1997) have argued that Redelmeier and Tibshirani's (1997a) study was too small, and hence did not have sufficient statistical power, to confidently

which is consistent with the conclusions of the large accident reports cited in the preceding chapter (see Shinar, 1978; Sussman, Bishop, Madnick, and Walter, 1985; Treat et al., 1977), is also consistent with the conclusion of a number of other studies (see Arthur, Barrett, and Alexander, 1991; Arthur and Strong, 1991; Goodman et al., 1999; Klauer et al., 2006; Langham, Hole, Edwards, and O'Neil, 2002; Larsen and Kines, 2002; McEvoy et al., 2007a; Murray, Ayres, Wood, and Humphrey, 2001; Violanti, 1997). For example, driver inattention to the roadway was found to be a contributing factor in 78 per cent of the crashes and 65 per cent of the near-misses captured in the recent naturalistic study involving 100 instrumented vehicles (see Klauer et al., 2006).

This is not, of course, to assert that the act of, say, dialling a number will not in-and-of-itself lead to a performance deficit. In fact, a number of studies have now shown that driving performance is often impaired when people have to manipulate their mobile phones (or other in-car equipment, such as the car radio; see also Chapter 3) at the same time (e.g., Briem and Hedman, 1995; Brookhuis, De Vries, and De Waard, 1991; Fuse, Matsunaga, Shidoji, Matsuki, and Umezaki, 2001; Graham and Carter, 2001; Haigney, Taylor, and Westerman, 2000; McKnight and McKnight, 1993; Patten et al., 2004; Salvucci, 2001; Salvucci and Macuga, 2002; Serafin, Wen, Paelke, and Green, 1993a, b; Stein, Parseghian, and Allen, 1987; Törnros and Bolling, 2005; though see also Mazzae et al., 2004). For example, in one study, Haigney and colleagues demonstrated that participants were significantly worse at lane-keeping in a driving simulator when they used a handheld rather than a hands-free phone (see also Brookhuis et al., 1991; Stein et al., 1987, for similar results). Meanwhile, Jenness, Lattanzio, O'Toole, Taylor, and Pax (2002) have shown that participants who initiated their phone calls using a voice-activated dialling system were better able to keep in lane, and made fewer glances away from the forward roadway, than those drivers who had to dial the numbers manually in their simulated driving study (involving participants playing a video game). Finally, another study conducted at the driving simulator at the TRL (Transport Research Laboratory) in Wokingham, England has shown that drivers who use a handheld phone also take longer to slow down than drivers who use a hands-free phone ('The mobile phone report', 2002; see Figure 2.1).

Support for the role of attention in driving-related accidents

Kahneman, Ben-Ishai, and Lotan (1973) conducted one of the first empirical studies to look for a possible link between a person's ability to selectively attend and their involvement in road traffic accidents. The researchers tested 39 Israeli bus drivers who had had at least two moderately severe accidents in the year preceding the study. They compared the performance of their drivers on a selective listening task to that of a control group of drivers who had either had none (or just one) accident while driving their bus professionally. Kahneman and his colleagues used a modified version of the selective listening task popularized by Colin Cherry

---

determine whether or not there was a difference in road safety between handheld and hands-free phones.

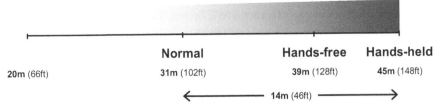

Figure 2.1 **Schematic diagram showing the distance that a car would travel before a driver would respond for a vehicle travelling at 70 mph**

(Figure redrawn and adapted from 'The mobile phone report', 2002)

(1953, 1954) and Donald Broadbent (e.g., 1958; see Driver, 2001, for a review). Pairs of words were presented to the bus drivers over headphones at a rate of 2 pairs of words per second (one to either ear). The drivers had to shadow (i.e., to repeat out loud) the words presented to one ear while trying to ignore the words presented simultaneously to their other ear. Periodically, an auditory beep was presented to one or other ear that indicated to the participants whether they should either switch to shadowing the other ear (if the beep was presented to the unshadowed ear), or else keep shadowing the same ear (if the beep was presented to the ear that was currently being shadowed).

Kahneman et al. (1973) found that the more accident-prone bus drivers made significantly more errors on the shadowing task (following the instruction to shadow the other ear) than did the control drivers. It was hypothesized that these errors highlighted the cost associated with the switching of a person's attention from one source of information to another.[3] It is, however, once again, important to note that the existence of an association between the number of traffic accidents that drivers had been involved in in the preceding year and their selective listening performance does not necessarily tell us anything about causality: for while it might be the case that attentional deficits give rise to an increased risk of having an accident (as Kahneman et al. suggested; see also Arthur, Barrett, and Alexander, 1991; Arthur and Doverspike, 1992; Avolio, Kroeck, and Panek, 1985; Mihal and Barrett, 1976), it seems equally plausible that having two or more accidents within a 12-month period might also affect one's concentration (see Korteling, 1990; Miller, 1970; Stuss, Stethem, Hugenholtz, Picton, Pivik, and Richard, 1989; van Zomeren and Deelman, 1976; Wood, 1988).

More convincing evidence concerning the potentially deleterious effect of talking (or using a mobile phone) while driving has come from research in applied psychology that has attempted to investigate more directly when, and how, performing an auditory task interferes with a person's driving performance. One of the first studies to have been published in this area detailed the seminal work of Ivan Brown, based in Cambridge, England. Brown and his colleagues conducted a number of the earliest studies that attempted to determine the cause of any dual-task

---

3 In fact, there was even talk of using this kind of selective attention task in order to select drivers/pilots to hire (see Arthur and Strong, 1994; Doverspike, Cellar, and Barrett, 1986; Gopher and Kahneman, 1971; Gopher, 1982).

interference while a driver was driving (see Brown, 1962, 1965; Brown and Poulton, 1961; Brown, Simmonds, and Tickner, 1967; Brown, Tickner, and Simmonds, 1969). Brown and his colleagues' studies are particularly significant because they constitute some of the earliest empirical research to have assessed driving performance while a person was actually behind the wheel of a car.

The 24 male drivers (median age of 41 years) in Brown et al.'s (1969) study had to drive a manual Austin A40 estate car on an airfield test track and judge whether the vehicle could be manoeuvred through a series of 20 openings that had been laid out on the track. Some of the openings were slightly wider than the car itself, while others were slightly narrower (in fact, the gaps varied in width from 3 inches narrower than the driven car to 9 inches wider in 3 inch steps). At the same time, the participants sometimes had to report on the logical consistency of a series of short sentences that were presented over the car radiophone (as such devices were known in those days). The loudspeaker was mounted directly in front of the driver who responded verbally using a headset, thus ensuring that the driver's hands were kept free. The drivers had to make speeded 'true' versus 'false' responses in response to short statements such as 'A follows B' – 'BA' presented from a mobile transmitter truck parked next to the text track.

The key result to emerge from Brown et al.'s (1969) study was that the performance of the auditory task impaired drivers' judgments of the gaps (especially the impossible gaps), but had little effect on their ability to steer through the possible gaps, although their driving was slowed significantly (by 6.6 per cent). The speed and accuracy of participants' responses on the sentence-checking task was also impaired when the participants were driving: in fact, there was a two-fold increase in the number of errors that they made.[4] Brown et al. suggested that the use of a phone while driving probably resulted in a perceptual or decisional impairment that was caused, at least in part, by the division of their drivers' attention between different sensory modalities (i.e., between their eyes and ears), and attentional-switching costs (cf. Kahneman et al., 1973). On the basis of their results, Brown et al. went on to conclude that talking was likely to have only a minimal effect on the more automatized of driving skills (such as steering), but that perception and decision-making abilities were critically impaired by drivers having to switch their attention repeatedly between their eyes and ears (see Spence, Nicholls, and Driver, 2001, for an estimate of the time taken by people to switch their attention between the auditory and visual modalities). As we shall see in the pages that follow, the majority of the subsequent research that has been conducted on the topic of talking while driving has provided support for many of Brown et al.'s early conclusions.

Dividing attention between eye and ear

Support for the claim that mobile phone use may adversely affect driving performance by distracting a driver's attention away from what he or she is looking at (i.e., from the visual modality; Spence and Ho, forthcoming; Strayer and Johnson, 2001) has

---

4   It is, however, worth noting that the participants were given no explicit instructions concerning how to prioritize the two tasks in this study.

come from a number of more recent studies (e.g., McCarley, Vais, Pringle, Kramer, Irwin, and Strayer, 2004: Strayer, Cooper, and Drews, 2004; Strayer and Drews, 2007; Strayer, Drews, and Johnston, 2003). For example, Strayer et al. (2004) reported an experiment in which they showed that having a conversation on a mobile phone with one of the experimenter's confederates impaired participants' recall of specific roadside signs, advertisements, cars, trucks, and pedestrians. Moreover, by monitoring their participants' eye movements, Strayer and his colleagues were able to show that the decrement in performance was caused by participants paying less attention to the information presented (no matter whether it was presented at fixation or not) than when they had no auditory task to perform. That is, the participants exhibited a reduced memory for visual stimuli, even for those stimuli that the participants happened to have fixated on (regardless of their relevance to driving; see also Langham et al., 2002).

It should, however, be noted that in order to assess their participants' visual memory, Strayer et al. (2004) presented them with a surprise two-alternative forced choice recognition memory task immediately after they had completed the simulated driving task. Consequently, it is not really possible to distinguish between drivers simply not 'seeing' the stimuli (the authors' preferred explanation for their own results), and drivers seeing the stimuli when they were presented but then rapidly forgetting about them (this distinction has been categorized by Wolfe, 1999, as one between 'inattentional blindness' and 'inattentional amnesia').

Over the last decade, a number of researchers have also used the change blindness paradigm in order to investigate drivers' visual perception (see Batchelder et al., 2003; Rensink, O'Regan, and Clark, 1997; Velichkovsky, Dornhoefer, Kopf, Helmert, and Joos, 2002). This research clearly shows that people can be 'blind' to (or at least be unable to consciously detect or recall) the appearance or disappearance of a variety of different visual stimuli if the onset of the change happens to be masked by a local visual transient (such as a mudsplash, or the flashing of an item on a head-up visual display projected onto the windscreen; O'Regan, Rensink, and Clark, 1999; see also Auvray, Gallace, Tan, and Spence, 2007; Tufano, 1997). Similarly, the occurrence of a global visual transient, such as caused by a person blinking, has also been shown to result in change blindness using driving scenes (Velichkovsky et al., 2002). Interestingly, research utilizing the change blindness paradigm has demonstrated that people are not only blind to visual changes that are peripheral to their driving task (such as the sudden appearance or disappearance of an advertising hoarding by the side of the road), but, perhaps more worryingly, that they are also apparently blind (or at least slower to respond) to important changes that may be taking place on the roadway itself, such as, for example, a change in the road markings (Rensink et al., 1997), or the sudden appearance of a cyclist on the roadway ahead (McCarley et al., 2004; Richard, Wright, Prime, Ee, Shimizu, and Vavrik, 2002; Velichkovsky et al., 2002; see also Langham et al., 2002).

Researchers have also used the change blindness paradigm in order to investigate whether participants pay less attention to what they can see (or are looking at) when they are simultaneously engaged in an auditory dual-task (see McCarley et al., 2004; Richard et al., 2002). For example, McCarley and his colleagues demonstrated visual change blindness in a study where participants had to spot the changes taking place in complex traffic scenes (once again, the flicker paradigm was used to mask the local transients that are typically elicited by change; though see Fernandez-Duque and Thornton, 2000; Rensink, 2004). Importantly, McCarley et al. found that their participants needed to make more saccades in order to successfully detect and identify the change if they happened to be talking at the same time (see also Richard et al., 2002).

Taken together, the available evidence from the laboratory-based research on the change blindness paradigm shows that a person's visual perceptual abilities can be significantly impaired by the simultaneous performance of an auditory secondary task (McCarley et al., 2004; Richard et al., 2002). In particular, it would seem that the distraction (or cognitive workload) associated with the performance of an auditory secondary task impairs people's ability to detect sudden changes in the visual scene (in terms of both the speed and accuracy of their change detection responses). Diverting attentional resources (or 'spare mental capacity', see Brown, 1962) toward the auditory modality (and/or toward the location from which the auditory stimuli are presented; see below) appears to degrade the processing of visual stimuli/information no matter whether it is presented at fixation or else out in the periphery (Strayer et al., 2003, 2004; see also Barkana, Zadok, Morad, and Avni, 2004; Jolicouer, 1999; Recarte and Nunes, 2000, 2003). Interestingly, the latest research to emerge from the Strayer and Drews laboratory (Strayer and Drews, 2007) has also shown reduced visual event-related potentials (ERPs) in drivers in response to specific visual stimuli when they are given a secondary auditory conversation task, thus supporting the genuinely perceptual nature of this impaired processing of visual stimuli (see also Beck, Rees, Frith, and Lavie, 2001; Fernandez-Duque and Thornton, 2000; Garcia-Larrea, Perchet, Perrin, and Amenedo, 2001).

Engaging in a simulated conversation also appears to decrease the efficiency of a person's oculomotor search (once again, as originally hinted at by Brown et al., 1969). Meanwhile, other research has also shown that drivers tend to look away from the roadway more when they are manually dialling a telephone number into their mobile phone as well (see Jenness et al., 2002). The latest research from Harbluk, Noy, Trbovich, and Eizenman (2007), who monitored the eye movements of a group of 21 drivers on the road while given either no secondary task, an easy cognitive task (involving the addition of two single digit numbers, e.g., '7 + 8'), or a hard secondary task (involving the addition of two double digit numbers instead, e.g., '74 + 38') performed in a hands-free manner. They found that performing the hard task resulted in drivers spending more of their time looking straight ahead, and less time looking away to more peripheral regions. The drivers in their study also looked at their instruments and mirrors less frequently in the hard cognitive task condition (see also Smith, Luke, Parkes, Burns, and Landsdown, 2005).

Talking while driving not only impairs oculomotor search (or visual scanning) and visual perception, but has now also been shown to impair (i.e., slow) people's performance in a wide range of different (driving-related) tasks (see Alm and Nilsson, 1995; Amado and Ulupinar, 2005; Beede and Kass, 2006; Brookhuis et al., 1991; Brookhuis, de Waard, and Mulder, 1994; Brown et al., 1969; Consiglio et al., 2003; Fuse et al., 2001; Hancock, Lesch, and Simmons, 2003; Horrey and Wickens, 2006; Irwin, Fitzgerald, and Berg, 2000; Lamble et al., 1999; Richard et al., 2002; Rosenbloom, 2006; Strayer and Drews, 2004, 2007; Strayer, Drews, and Crouch, 2006; 'The mobile phone report', 2002; Treffner and Barrett, 2004; though see also Rakauskas, Gugerty, and Ward, 2004). Such results are entirely consistent with the view that drivers have only a limited pool of attentional resources (Horrey, forthcoming), that is (at least to a large extent) shared between the processing of the information that is available in different sensory modalities (see Spence and Driver, 1997a; Spence and Ho, forthcoming; though see Wickens, 1984, 1992, for a contrasting viewpoint; Hancock, Oron-Gilad, and Szalma, 2007). In fact, a meta-analysis of 23 different studies on the effects of using a mobile phone while driving recently carried out by Horrey and Wickens (2006) suggested that the reaction time deficit was on average around 130 ms[5] (see also Rumar, 1990). Interestingly, a more recent meta-analysis of 33 studies carried out by Caird, Willness, Steel, and Scialfa (forthcoming) has suggested a somewhat higher figure of around 250 ms instead. Finally, it is worth noting that using a phone while driving has also been reported to result in an increase in mental workload, as, for example, measured by objective measures such as increased heart rate (Haigney et al., 2000) or pupillary dilation (Recarte and Nunes, 2000), and by people's subjective responses to workload questionnaires, such as provided by the NASA-TLX (Alm and Nilsson, 1995; Brookhuis et al., 1991; Horberry, Anderson, Regan, Triggs, and Brown, 2006; Rakauskas et al., 2004; Tokunaga et al., 2004; Waugh, Glumm, Kilduff, Tauson, Smyth, and Pillalamarri, 2000; see also Törnros and Bolling, 2006).

The costs associated with dividing attention spatially

It seems unlikely that the costs associated with using a mobile phone while driving have a unitary cause. One factor that has not been discussed thus far in this chapter, but which may contribute to the dual-task decrement relates to the requirement for drivers who are conversing on mobile phones to divide their attention spatially between two or more different locations. A large body of laboratory-based research has shown that people find it harder (i.e., their performance deteriorates) when they have to direct their auditory and visual attention in different directions rather than focusing all of their attention in the same direction (Driver and Spence, 1994; Eimer, 1999; Spence and Driver, 1996; see Driver and Spence, 2004, for a recent review).

Spence and Read (2003) conducted a series of experiments in the Leeds Advanced Driving Simulator (LADS) in order to investigate whether the same constraints on dividing attention spatially would also be observed when drivers performed a demanding auditory task whilst at the wheel. The participants in their study had to sit in a manual Rover 216 GTI car with 130° visual field of view and realistic road and car noise. The participants either had to perform a selective listening (i.e., shadowing) task while the vehicle was parked in a lay-by, or else to perform the shadowing task while they were driving along a visually demanding suburban road network. The participants' auditory task involved shadowing one of two simultaneously-

---

5   Note that this matches the time required to switch attention between audition and vision in laboratory studies when stimuli in two modalities are presented from different spatial positions (Spence, Nicholls, and Driver, 2001).

presented streams of auditory speech (as in a typical shadowing task). The relevant speech stream was either presented from the side, or, in other blocks, from directly in front of the participant, while the irrelevant (i.e., distracting) auditory stream was always presented from a loudspeaker cone placed midway between the other two loudspeakers to the participant's front-left (see Figure 2.2). The participants' task was made more difficult by the fact that the words were presented in background white noise to mask them (i.e., the perceptual load of the participants' auditory and visual tasks, see below, was kept relatively high; cf. Lavie, 2005; Spence, Ranson, and Driver, 2000b).

The results of Spence and Read's (2003) study showed that their participants found it significantly easier to listen to what the relevant speaker was saying when the auditory stimuli were presented from the front rather than from the side. This may reflect the well-known frontal speech advantage (Hublet, Morais, and Bertelson, 1976, 1977; Morais, 1978). However, more importantly for present purposes, was Spence and Read's finding that this frontal speech advantage was more pronounced when the participants had to drive the vehicle in the simulator at the same time. These results therefore provide empirical support for the claim that one factor that may contribute to making it more difficult to hold a mobile phone conversation while driving is related to the difficulty people have in dividing their attention spatially between different locations (cf. Fagioli and Ferlazzo, 2006; Jenness et al., 2002). Such results have even led some commentators to put forward the idea of introducing 'talking windscreens' to ensure that a driver's auditory and visual attention will typically be focused in the same direction (i.e., toward the front; cf. Stein et al., 1987; Wikman et al., 1998). However, it should also be pointed out here that having a stream of auditory speech emanating from the direction in which a

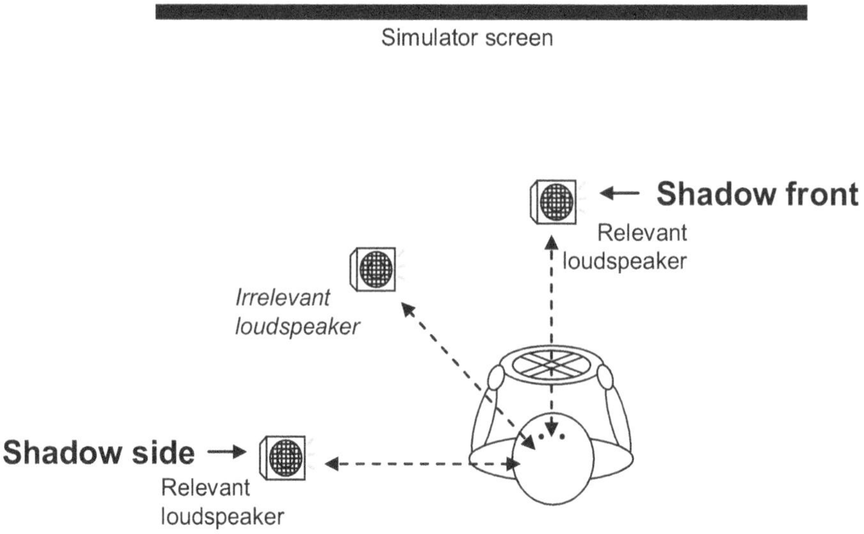

Figure 2.2   Schematic diagram of the experimental set-up used in Spence and Read's (2003) driving simulator study of performing an auditory task while driving

person is looking will also make it harder to ignore that speech stream when one is trying to concentrate visually on the forward roadway (see Spence et al., 2000b, on this point). Furthermore, it is also important to remember that the analysis of the driving performance measures in Spence and Read's study failed to reveal any evidence that their participants' driving was adversely affected by changing the relative direction from which the relevant (i.e., to-be-shadowed) auditory speech stream was presented (perhaps hinting at the successful prioritization of the driving task by the participants in this study; see below).

Impaired decision making

Horswill and McKenna (1999) conducted a laboratory-based study that provided support for Brown et al.'s (1969) early suggestion concerning the detrimental effect of auditory task performance on a driver's decision making abilities. They assessed the risk-taking behaviours of 121 undergraduate drivers using a video-based task involving the judgment of close-following, gap acceptance, and hazard perception. In the close-following task, participants watched video footage taken through the windscreen of a car that gradually approached the rear of another car on a motorway. The participants had to press a button as soon as the car reached the distance that they would normally travel behind a car on the road, and to press the button again when they started to feel uncomfortably close to the lead vehicle in the video. Meanwhile, in the gap acceptance task, the participants viewed video footage of a road junction taken from the point of view of a driver who was waiting to turn left and join a stream of traffic. The participants were instructed to respond whenever they would have pulled out into the stream of traffic. Finally, in the hazard perception task, the participants had to press a button as rapidly as possible when they saw a potentially hazardous situation in the video. These tasks were either performed in isolation, or while the participants simultaneously performed a verbal task. A list of letters was read out (at a rate of 1 per second), and the participants had to say 'yes' whenever the letter 'K' was presented, otherwise they had to say 'no'.

The results showed that the participants in this study took more risks (i.e., they behaved more dangerously in all three driving decision-making tasks) in the dual-task conditions than in the single task condition. What is more, their performance on the verbal task was also significantly worse than that of a control group of participants who only performed the verbal task. Horswill and McKenna (1999) went on to argue that their results demonstrated that risk-taking judgments are not automated. The shift of a driver's visual attention to the roadway in the event of a driving incident was captured in Horswill and McKenna's results by the temporary drop in their participants' ability to perform the verbal task that occurred just at the moment when they were making their hazard perception decisions (i.e., when danger was detected on the road in the video). This result therefore adds support to attention-switching accounts of the dual-task decrement associated with using a mobile phone while driving (Brown et al., 1969; Kahneman et al., 1973; see also Elander, West, and French, 1993). However, it is important to stress the artificiality of Horswill and McKenna's experimental paradigm, which involved relatively inexperienced drivers making decisions about events seen in a video presented in a psychology laboratory while being instructed to prioritize their performance of the verbal task. Another potential limitation relates to the fact that the mean age of the participants in their study was less than 21 years, for it is well known that younger drivers are much more accident-prone than older drivers (e.g., see Doherty, Andrey, and MacGregor, 1998; Evans, 1991). On the other hand, though, it is perhaps also worth noting that younger drivers are much more likely to use a mobile phone while driving than older drivers (Royal, 2003; Stutts et al., 2001; 'The mobile phone report', 2002).

Interim conclusions

Over the 40 years or so since Ivan Brown and his colleagues conducted their seminal research in the area, many studies have been designed to address the question of just when, and how, the use of a mobile phone interferes with driving performance. The data from the driving studies reported thus far in this chapter reveal that performing an auditory task while driving impairs people's risk-taking judgments or behaviours, such as demonstrated by the results of studies highlighting people's ability to judge gaps and to make close following judgments (Brown et al., 1969; Horswill and McKenna, 1999; see also Bowditch, 2001; Cooper et al., 2003). However, it is important to note that not all driving tasks are affected to the same extent by talking on a mobile phone. For instance, steering and mirror checking seem to be relatively unaffected by the performance of an auditory secondary task (see Brookhuis et al., 1991; Brown et al., 1969).

The majority of the studies that have been reported thus far in the present chapter can, however, be criticized in terms of the ecological validity of the auditory (and, in some cases, visual) tasks that have been utilized (cf. Horrey and Wickens, 2006; Shinar, Tractinsky, and Compton, 2005). The auditory tasks that have been used have varied from selective shadowing tasks (Kahneman et al., 1973; Spence and Read, 2003) to speech monitoring (Horswill and McKenna, 1999), and from logical reasoning studies (Brown et al., 1969) to maths problem solving (Harbluk et al., 2007; McKnight and McKnight, 1993; Shinar et al., 2005; Treffner and Barrett, 2004) and tasks that have tapped into auditory working memory (Alm and Nilsson, 1994; Brown, 1962; Brown and Poulton, 1961; Hancock, Simmons, Hashemi, Howarth, and Ranney, 1999; Langham et al., 2002; see Wetherell, 1981, for an early review of the literature).

However, while such carefully structured verbal tasks clearly have certain advantages (i.e., in terms of maintaining a constant level of task difficulty; Fuse et al., 2001; Shinar et al., 2005), it is also apparent that in order to provide greater external validity to the research conclusions that are drawn, researchers really ought to implement more naturalistic conversational tasks in their experimental studies (Haigney and Westerman, 2001). It would seem intuitively obvious that conversational tasks would presumably offer the closest approximation to the demands of holding a mobile phone conversation while engaged in real driving (Haigney and Westerman, 2001; McCarley et al., 2004). Indeed, one legitimate question to ask at this point then concerns how (or whether) the dual-task decrements that have been identified thus far would change if people were engaged in an actual conversation rather than one of these other less realistic (or less ecologically-valid) auditory tasks (see Horrey and Wickens, 2006). It is in this context that the research of David Strayer and his colleagues at The University of Utah (e.g., Strayer et al., 2003, 2004; Strayer and Johnston, 2001) really stands out. For, over the last seven years, these researchers have conducted a large number of studies in which the participants have been engaged in reasonably naturalistic conversation (see also 'The mobile phone report', 2002). While the visual driving tasks used in their original research can be criticized for lacking ecological validity, this aspect of their experiments has improved substantially in their more recent published research in this area.

In their original laboratory-based research studies, Strayer and Johnston (2001) addressed the question of whether engaging in a naturalistic conversation with a confederate of the experimenter would result in a dual-task decrement on a laboratory-based driving task that was any more (or less) severe than that identified when another, less naturalistic, auditory task was used. In particular, the 48 young participants (mean age of 21 years) in their study were either engaged in a conversation concerning current affairs by the experimenter's confederate (involving a discussion either of the Clinton impeachment or the Salt Lake City Olympic Committee bribery scandal), or else they had to listen to a radio station of their choice instead. The participants also had to perform a laboratory-based visual pursuit tracking task. In this task, participants used a joystick to move a cursor on a computer display in order to try and keep it aligned with a target that moved unpredictably across the screen in front of them. The participants also had to make speeded button-press 'braking' responses on the joystick to occasionally-presented red stop signals flashed on the computer display every 10–20 seconds. Green lights were also presented equiprobably, but the participants were instructed to make no response to these 'catch' stimuli. Strayer and Johnston's results showed that those participants who were engaged in unconstrained conversation missed significantly more stop signals (4 per cent fewer stop signals were detected overall; equating to a twofold increase), and they also responded more slowly to those stop signals that they did detect. Interestingly, Strayer and Johnston also found that talking was found to be more detrimental to driving performance than listening.

In a second experiment, Strayer and Johnston (2001) went on to try and determine precisely what aspect of conversing it was that was causing the impairment on the simulated driving task. The participants now either simply had to shadow a list of words they heard over a handheld mobile phone, or else they had to generate a new word that began with the last letter of each of the words that they heard (word generation task). Strayer and Johnston varied the difficulty of the tracking task by varying the predictability of the movement of the central target that the participants had to try and follow with the joystick (note that the peripheral target detection task was no longer presented in this second experiment). Strayer and Johnston's results showed that while neither task interfered with performance (as now measured by the RMS error on the tracking task) on the easy tracking task, significant interference was observed in the word generation task (but not in the speech shadowing task) when the participants had to perform the more difficult tracking task.

Although these results might be taken to demonstrate that the generation of speech causes more interference with driving performance than the act of speaking itself (as presumably indexed by the performance of the shadowing task; see also McCarley et al., 2004), it should be noted that Strayer and Johnston (2001) themselves argued that it is the *difficulty* of the tasks involved that will determine how difficult people find it to perform a driving task and an auditory secondary task at the same time. In this regard, it is worth noting that other research has shown that shadowing speech can lead to a slowing of participants' responses to critical visual events on the windscreen of as much as 370 ms when compared to participants who merely had to listen to the same message (i.e., without shadowing; see Fuse et al., 2001).

Speech production versus speech comprehension

Evidence in support of the claim that there may not be any qualitative difference in the difficulty of comprehending versus generating speech while driving comes from a recent series of experiments by Kubose, Bock, Dell, Garnsey, Kramer, and Mayhugh (2006). The participants in this driving simulator study performed a speech task and a simulated driving task under both single- and dual-task conditions. The researchers compared the effects of speech production to those of speech comprehension. Crucially, and in contrast to the study of Strayer and Johnston (2001), the linguistic complexity of the two auditory tasks was carefully matched. Kubose et al. found that the drivers exhibited more variable performance – in terms of their driving speed in one experiment (when the participants were trying to maintain constant speed of 55 mph); and in terms of their ability to maintain a constant headway (when the participants had to try and follow behind an erratic lead vehicle) in another experiment – when either speaking or listening to speech than when only performing the simulated driving task. Importantly, however, the two linguistic tasks had very similar effect on the majority of driving measures that were assessed.

Kubose et al.'s (2006) results therefore support the idea that there is no fundamental difference between the dual-task costs associated with generating versus comprehending speech. Instead, the detrimental effect on driving of speech comprehension actually appears to be about the same as for speech generation, at least when the tasks are matched for their linguistic difficulty. Taken together, these results are consistent with the findings of several other studies in suggesting that the amount of interference elicited by a mobile phone task while driving will critically depend on the attentional or cognitive demands of each of the concurrently-performed tasks (Doherty, Andrey, and MacGregor, 1998; Graham and Carter, 2001; Haigney and Westerman, 2001; Horrey and Wickens, 2006; Kantowitz, Hanowski, and Tijerina, 1996; Nunes and Recarte, 2002; Recarte and Nunes, 2003; Strayer and Johnston, 2001). Thus, the more complex (and/or emotionally demanding; see Shinar et al., 2005) the dialogue, the larger the performance decrement in simulated driving is likely to be (Amado and Ulupinar, 2005; Boase, Hannigan, and Porter, 1988; Briem and Hedman, 1995; McKnight and McKnight, 1993; Patten et al., 2004; though see also Irwin et al., 2000).

Taken together, the results of this research would seem to be most parsimoniously accounted for in terms of Nilli Lavie's perceptual load theory (e.g., Lavie, 1995, 2005, for a review). Lavie put forward her theory in order to try and account for why attentional selection sometimes appears to take place early in information processing, while at other times the attentional gating appeared to operate much later. According to Lavie, the perceptual load (or difficulty) of a task has a critical impact on whether attentional selection takes place early or late (see also Brown, 1962, for an early discussion of capacity and driving). The research reviewed in the last two sections of this chapter would appear to be consistent with the claim that the harder (or more perceptually demanding) a task is, the more performance on another secondary task will be seen to deteriorate due to capacity constraints in attentional processing.

Lavie and her colleagues have also shown that loading a person's working memory impairs the ability to keep focused on a relevant task and to ignore irrelevant task stimulus (e.g., de Fockert, Rees, Frith, and Lavie, 2001; see also Conway, Cowan, and Bunting, 2001; Yi, Woodman, Widders, Marois, and Chun, 2004). In fact, it is interesting to highlight the similar experimental methodologies now being used by researchers studying attentional selection in the laboratory (such as de Fockert et al.), and by applied researchers interested in dual-task limitations while driving on the test track (Hancock et al., 2003). In both of the aforementioned studies, participants (or drivers) are given a random string of digits (or telephone number) for subsequent recall. The investigators then evaluate how this concurrent memory load interferes with performance in another task – a visual selective attention task in de Fockert et al.'s study, and a driving task in Hancock et al.'s study. It is interesting to note here that many of the dual-task driving studies that have been reviewed in this chapter have involved loading a person or driver's short-term memory (see Alm and Nilsson, 1994; Brown, 1962; Brown and Poulton, 1961; Hancock et al., 1999, 2003; Harbluk et al., 2007; Langham et al., 2002).

It is our belief that the recent experimental and theoretical work on the concept of perceptual load offers a more promising framework for understanding why performing an auditory task may interfere with driving performance than has been offered by more traditional workload models of human information processing such as Wickens' multiple resource theory (see Chapter 1; though see also Hancock et al., 2007; Horrey and Wickens, 2006; Strayer and Johnston, 2001; Wickens, 2002). However, given that, at present, we only have an operational definition of the 'perceptual load' of a given task (Lavie and Tsal, 1994), further progress would be facilitated in this area if an independent measure of this could be developed.

Talking to a passenger versus talking on the mobile phone

Many people have raised the question of whether or not using a mobile phone while driving is actually any more dangerous than talking to a passenger (Laberge, Scialfa, White, and Caird, 2004; McKnight and McKnight, 1993; Haigney and Westerman, 2001; Parkes, 1991b). In fact, early research suggested that mobile phone conversations tend to be perceived subjectively as more effortful than conversations with a passenger (Kames, 1978). Certainly, the quality of the speech stream is not necessarily always that good when conversing via a mobile phone (Kawano, Iwaki, Azuma, Moriwaki, and Hamada, 2005), and this may be one of the factors helping to account for the increased 'psychological distance' that drivers often report when talking to someone on a mobile phone (Fairclough et al., 1993). Parkes (1991a) has reported that drivers performed significantly less well on a verbal task involving the presentation of questions taken from intelligence tests than when questions were delivered by a passenger. It has also been suggested that the nature of the conversation that drivers typically have with their passengers is likely to be qualitatively different from the kind of conversations that people have with others on the mobile phone (Drews, Pasupathi, and Strayer, 2004; Gugerty, Rando, Rakauskas, Brooks, and Olson, 2003; Horrey and Wickens, 2006; Waugh et al., 2000).[6] It has, for example, been suggested that people may sometimes engage in more emotionally- and/or intellectually-demanding conversations when speaking to someone over the mobile phone than when conversing with their passengers (cf. Horrey and Wickens, 2006). So, for example, Haigney and Westerman (2001) report on the results of one survey of in-car conversations that revealed that 65 per cent of conversations taking place on a mobile phone involved 'intense verbal negotiation'. Interestingly, interacting with a passenger is cited more frequently than mobile phone use as playing a causal role in distraction-related car crashes (Stutts, Reinfurt, and Rodgman, 2001).

Researchers have suggested that car passengers may pace (or regulate) their conversation with the driver in accordance with the current driving conditions (Haigney and Westerman, 2001; McKnight and McKnight, 1993). More specifically, the claim is that passengers tend to stop speaking when the driving conditions become more taxing (such as when driving on congested urban roads; see Gugerty, Rakauskas, and Brooks, 2004; Parkes, 1991b; 'The mobile phone report', 2002; though see Laberge et al., 2004, for contradictory evidence). Given that the person at the other end of the mobile phone is not aware of what is happening on the roadway he or she may consequently put the driver at higher risk than passengers do by demanding attention when the driver needs it to cope with a critical situation in the traffic flow (see Piersma, 1993; Summala, 1997). However, empirical evidence on this question has shown that no such conversational pacing is seen when drivers talk with someone on their mobile phones (see Crundall, Bains, Chapman, and Underwood, 2005; Gugerty et al., 2004; Nunes and Recarte, 2002).

Crundall et al. (2005) compared the number of utterances made by drivers and the people they were speaking to, when those people were either passengers in the car or else at the other end of a hands-free mobile phone. The conversational task involved the two people competing to win points. The results of this on-road study showed that in-car conversation was suppressed when the driving conditions became more demanding (i.e., on urban roads, as compared to rural roads or dual-carriageway driving). No such suppression or modulation of conversation was observed when the other person was at the end of a mobile phone. In fact, if anything, drivers made

---

6   Though note that it seems very plausible that there has been a change in the content of conversations over the years as mobile phones have changed from being the play-thing of business executives to the ubiquitous device carried (and used) by most young people (cf. Brown et al., 1969; Sundeen, 2006).

more utterances under such conditions than they did during 'normal' conversation (see also Alibali, Heath, and Myers, 2001; Gugerty et al., 2004).

Sagberg (2001) conducted a large questionnaire-based study involving 9000 Norwegian car drivers who replied to a postal survey (31 per cent return rate), concerning a recent crash that they had been involved in. The drivers' questionnaire responses indicated that approximately 10 per cent of their crashes were attributed to drivers having been conversing with a passenger at the time of the accident. By contrast, in-vehicle factors (including the use of a mobile phone) were mentioned as playing a causal role in just 6.9 per cent of crashes. Meanwhile, another study of 1367 drivers who visited hospital following a crash in Perth, Australia, found that 11 per cent of those interviewed cited the fact that they were talking with a passenger as a contributing factor to their accident (McEvoy, Stevenson, and Woodward, 2007a).

Recently, McEvoy et al. (2007b) conducted an epidemiological study of 274 drivers who visited a hospital following a motor vehicle crash and more than 1000 other control drivers. This case-crossover study demonstrated that the risk of a person having an accident while driving increases as a function of the number of passengers in the vehicle. While carrying a passenger resulted in an increase in the likelihood of having a motor vehicle crash resulting in hospital attendance of approximately 60 per cent, the risk actually doubled whenever two or more passengers were in the car (see also Doherty et al., 1998; McEvoy et al., 2007b; cf. Haigney and Westerman, 2001). This increase in risk compared to a fourfold increased risk of crashing associated with drivers using a mobile phone in the 5 minute period preceding the crash.

McEvoy et al.'s (2007b) results therefore show that while carrying passengers substantially increases the risk of accident, the problem is not as severe as that associated with the use of a mobile phone (see also Hunton and Rose, 2005, for similar results). However, while the risk associated with carrying passengers is lower than that associated with mobile phone use, it is worth remembering that it is nevertheless likely to be implicated in a greater number of vehicular crashes given the higher background prevalence of transporting passengers (see also Stutts et al., 2001). Furthermore, it should also be borne in mind that different sections of the population may be differentially affected by the presence of passengers in the car. For example, Doherty et al. (1998) have reported that the presence of passengers in a vehicle tended to be more detrimental for driving of younger drivers (16–19 years) than for older drivers (i.e., those aged 20–59 years; see also McEvoy et al., 2007a; Williams, Ferguson, and Wells, 2005).

The beneficial effects of mobile phones in vehicles

Finally, it should be noted that while this chapter (and the majority of the empirical research that has been conducted to date) has tended to focus on the deleterious effects of using a mobile phone while driving, it is important not to forget that talking while driving can also help to reduce boredom (this is especially important given the large number of accidents caused by drivers falling asleep at the wheel; see Maycock, 1996; Reyner and Horne, 1998; Sagberg, 1999). Obviously, any beneficial effects of conversing may become more apparent the longer a driver has been on the road (Drory, 1985; VCU, 2003). In this regard, it is worth highlighting the relatively short duration over which participants had to drive in the majority of studies in this area (see Brown et al., 1967; Drory, 1985, for exceptions). Speaking has also been shown to increase people's overall level of arousal, at least under certain conditions (Kinsbourne and Cooke, 1971; Liu, 2003; Mikkonen and Backman, 1988). Such an increase in alertness may help to explain the seemingly paradoxical finding reported in several of the studies reviewed in this chapter that performing an auditory secondary task can, on occasion, actually lead to an improvement in at least certain aspects of driving performance (see Briem and Hedman, 1995; Hove, Gibbs, and Caird, 2000; Kubose et al. 2006; Mikkonen and Backman, 1988; Strayer and Johnston, 2001).

It is also important to consider the possibility that drivers may actually take countermeasures when talking while driving, such as lowering their speed, and/ or increasing the following distance (Haigney et al., 2000; though see Alm and Nilsson, 1995; Boase et al., 1988; Cnossen, Meijman, and Rothengatter, 2004). Such compensatory activities might be expected to be more obvious in on-road studies where there are very real dangers associated with driving unsafely. By contrast, drivers may prioritize their tasks differently when in a driving simulator or when taking part in a laboratory-based driving study, where there are no real risks associated with dangerous driving (see Dressel and Atchley, forthcoming; Horrey and Wickens, 2006; Murray et al., 2001; Parkes, 1991b; Reed and Green, 1999; on this topic). However, as Haigney and Westerman (2001) note, there are serious ethical concerns regarding the use of on-road studies. They raise the important point that it is currently unclear who would be legally responsible, the experimenter or the participant, should a participant have a crash while driving in an experimental study. It is also worth pointing out that drivers are likely to be able to learn to combine dual-tasking with practice. Most studies have, however, involved relatively short experiments. In one of the few studies to look at this issue directly, Shinar et al. (2005) found that participants ability to complete hands-free phone tasks while in a simulator task improved quite substantially the course of five experimental sessions (see also Brookhuis et al., 1991, for similar results).

It is also worth remembering that mobile phones can also serve a very important safety function in the event of an accident (Redelmeier and Tibshirani, 1997a; Wortham, 1997), and can also relieve anxiety about unavoidable delays, and even potentially be life-saving in an emergency (i.e., after an accident). Finally, some economists have even tried to justify the continued use of mobile phones in cars in terms of their positive impact on business productivity (see Hahn et al., 2000), suggesting that while banning mobile phone use in cars would likely reduce the number of crashes, it would also result in an anticipated annual loss of $20 billion (see also Redelmeier and Weinsten, 1999; though see also Cohen and Graham, 2003, for a revised estimate).

Conclusions

The evidence reviewed in this chapter shows that using a mobile phone while driving normally results in a marked dual-task decrement (see also Just, Keller, and Cynkar, 2008). However, contrary to what one might believe on the basis of current legislation (at least in the U.K. and in certain states in North America), this difficulty does not primarily appear to be associated with manual interference, since the majority of studies have shown little, if any, difference between people's driving performance when using a handheld versus a hands-free mobile phone (e.g., Redelmeier and Tibshirani, 1997a; Strayer and Johnston, 2001). Instead the difficulty appears to be more attentional in nature (cf. Klauwer et al., 2006; Shinar, 1978; Treat et al., 1977). Research has shown that one problem seems to be related to the difficulty that people have in shifting their attention from one sensory modality to another (Spence et al., 2001), such as from concentrating upon the voice that they are trying to listen to back to visually inspecting the road ahead (Brown et al., 1969; Strayer, Drews, and Johnston, 2003; Strayer and Johnston, 2001). People (including drivers) also find it difficult to divide their attention spatially, such as when trying to listen in one direction while trying to look in another (see Driver and Spence, 2004; Spence and Read, 2003; see also Fagioli and Ferlazzo, 2006).

Overall, the attentional problems associated with talking while driving would seem to more adversely affect a driver's risk-taking and decision making (e.g., Brown et al., 1969; Cooper et al., 2003; Horswill and McKenna, 1999) than it does the more automatic aspects of driving performance, such as maintaining lane position and following distance (Briem and Hedman, 1995; Brookhuis et al., 1991; though see Kubose et al., 2006; Rakauskas et al., 2004). Horrey and Wickens (2006) draw a distinction here between continuous and discrete task performance, with the performance of an auditory secondary task having a more detrimental effect on the latter (though see Rakauskas et al., 2004).

We have focused, in this chapter, on the attentional limitations associated with one specific form of in-car technology – the mobile phone. However, it is our belief that the constraints and limitations identified here would also adversely affect a person's ability to drive while using other forms of in-car device (such as the satellite navigation, internet, speech-based email; see Chapter 1). Given that perhaps the most serious problem for drivers relates to inattention (be that being distraction, as in the research reviewed in this chapter, or drowsiness, as we will see in the next chapter) and not simply any manual problems associated with physically interacting with the phone, it is clear that warning signals need to be designed in such a way that they are capable of automatically capturing a driver's spatial attention and redirecting it toward the road as rapidly as possible.

Chapter 3

Driven to Listen

Introduction

One of the most common non-essential activities that people engage in while driving is listening to the radio (e.g., Bull, 2005; North, Hargreaves, and Hargreaves, 2004; Sloboda, 1999; Sloboda, O'Neill, and Vivaldi, 2001). Although research has shown that listening to the radio, or music in general, can affect people's mood and other aspects of their cognitive performance in both a positive and negative manner (see e.g., Krumhansl, 2002), only limited efforts have been made to extend these findings to the domain of driving (see Dibben and Williamson, 2007, for a recent survey). In particular, few studies have attempted to understand whether any systematic conclusions or generalizations can be drawn concerning the effect of listening to the radio on driver performance, and above all, on car accident rates.

At the same time, many researchers have reported on the detrimental effects on driving performance when drivers engage in other forms of dual-tasking, such as talking on the mobile phone (see Chapter 2; Horrey and Wickens, 2006; Stutts et al., 2005). This impairment in dual-task performance is thought to reflect the capacity limitations on human information-processing (see Chapter 1). Given that certain of the limitations derived from empirical studies of talking on the mobile phone while driving seem equally applicable to the distraction caused by listening to the car radio (though see Strayer and Johnston, 2001), the most obvious prediction would be that listening to the radio should impair driving performance under at least certain conditions. Indeed, a laboratory-based driving study by Jäncke, Musial, Vogt, and Kalveram (1994) showed that listening to the radio (a radio broadcast consisting of both music and news) resulted in a greater mean deviation from the centre of the road. However, as shown in the preceeding chapter, findings from other, more recent laboratory-based research on the concept of perceptual load would appear to suggest that the answer may not be quite so simple.

Given the increasing provision of auditory warning signals now designed to compensate for problems of visual overload during driving (see Belz, Robinson, and Casali, 1999; Ho and Spence, 2005a), it is particularly important to consider the safety implications of listening to the radio on driver performance. Indeed, it has been suggested that the sound of the car radio or other in-car entertainment system may potentially mask any critical auditory warning signals that happen to be presented in the vehicle (see Slawinski and MacNeil, 2002). One certainly has to wonder whether auditory warning signals could ever be effective amongst certain younger drivers, given the results of a study by Ramsey and Simmons (1993) in North America that documented young males in Stanford, California playing their car stereos at an ear-splitting 85–130 dB.

A similar problem has also been documented at the other end of the age spectrum. In particular, Slawinski and MacNeil (2002) have reported that many older drivers (aged 65–85 years of age in their study) find it difficult to hear external noises, such as the sound of sirens and other warning signals over the background road noise, especially when that background noise is combined with the sound of the car radio. These findings led Slawinski and MacNeil to conclude that playing music while on the road may potentially increase the accident risk amongst older drivers. Any such detrimental effects of listening to music while driving need to be weighed up against the potential beneficial effects that listening to music may have, such as keeping drowsy drivers awake (see Reyner and Horne, 1998; though see also Oron-Gilad, Ronen, and Shinar, 2008; Oron-Gilad and Shinar, 2000). In fact, research from Sagberg (2001) suggests that both the radio and CD players cause more accidents than using the mobile phone while driving (though note the differing incidence of these two activities).

It has been argued that the act of listening to music is often processed at a subconscious level (see Walker, 1979) that may influence drivers' behaviour without them necessarily being aware of it. Indeed, it would seem particularly challenging for investigators (or researchers) to collect accident data that can be attributed solely (or even partially) to a driver having being listening to the radio at the time of an accident (cf. Sagberg, 2001). Given this fact, it is important to evaluate the literature on the possible influence of listening to the radio on driver performance in order to better assess the safety implications of in-car entertainment systems.

Tuning in to the radio is commonly used by drowsy drivers as a means of attempting to suppress sleepiness (Maycock, 1996; Reyner and Horne, 1998). When considering the example of driving, it could be argued that some form of auditory stimulation may be beneficial in order for drivers to maintain an optimal level of arousal under monotonous driving conditions, such as, for example, when driving on a quiet motorway. It is only by gaining a better understanding of the effect of auditory stimuli (both music and speech) on human performance while driving, that people can be better informed of the potential risks and benefits of engaging in this ubiquitous activity (e.g., see Beh and Hirst, 1999; Smith, 1961; Wilkins and Acton, 1982). In this chapter, we focus primarily on the effects of passively listening to the radio while driving, although it should, of course, be noted that actively interacting with the radio (such as when tuning into a new station) have been studied elsewhere (e.g., Horberry, Anderson, Regan, Triggs, and Brown, 2006; Kames, 1978; Jordan and Johnson, 1993; Mouloua et al., 2003; Stein et al., 1987; Wikman, Nieminen, and Summala, 1998).

Wikman et al. (1998) conducted an on-road study of both experienced (i.e., those with a life-time driving experience of more than 50,000 km) and inexperienced car drivers (those with a life-time driving experience of less than 15,000 km). They found that when the less experienced drivers had to change a radio cassette or else tune the radio into a 'soft music' radio station, the lateral displacement of the car increased. Briem and Hedman (1995) have also investigated the effects of radio tuning on driving performance (see also Mouloua, Hancock, Rinalducci, and Brill, 2003; Stein et al., 1987).

Driver distraction

One of the major safety concerns regarding the use of in-car stereos while driving relates to driver distraction (Jordan and Johnson, 1993). This can be further sub-divided into the attentional consequences induced by the act of listening per se (i.e., cognitive demands) and the physical demands associated with the drivers' interaction with the hardware interface itself. As has been pointed out already, driver inattention has been identified as one of the leading causes of car accidents (Gibson and Crooks, 1938; Treat et al., 1977), with investigators reporting that a surprisingly high percentage (26–56 per cent) of all car accidents can be attributed to attentional and/or information-processing failures rather than to a lack of driving skill (Klauer et al., 2006; Shinar, 1978; Oron-Gilad et al., 2008).

Indeed, some traffic experts have drawn a link between low levels of driver arousal and car accidents. For example, Graham-Rowe (2001) estimated that as many as 10 per cent of all road fatalities may be caused by drivers falling asleep at the wheel, with others suggesting that one-third of all road accidents involved drowsy drivers (Sample, 2001; see also Maycock, 1996; Horne and Reyner, 1995).

Music, arousal and mood

Aside from its potential role in helping to keep drowsy drivers from falling asleep, the tempo of music also appears to modulate the speed at which a car is driven. In fact, the tempo of the music that people listen to has been shown to influence their performance in a variety of different tasks (e.g., Karageorghis, Jones, and Low, 2006). In the case of driving, it has been argued that the tempo of the music may once again induce a change in a driver's mood. For instance, Brodsky (2002) reported that increasing the tempo of music led to increased driving speed, virtual traffic violations, disregarded red traffic lights, lane crossings, and collisions in a video driving game study. However, given the lack of any actual physical danger associated with the participants' performance in such driving games, Brodsky's findings can only be taken as suggestive with regards to the claim that listening to fast music will increase their risk of having an accident in real driving situations. A similar criticism can also be levelled at the results of another laboratory-based driving study reported by North and Hargreaves (1999).

In one of the earliest studies to have investigated the effects of listening to music on drivers' control performance, Konz and McDougal (1968) reported that drivers drove faster when listening to slow music or Tijuana Brass (fast) music than when no music was presented. Konz and McDougal argued that this result reflected a change in mood and/or arousal induced by listening to the music, even though the tempo of the music itself did not seem to modulate driver performance. Increased active control activities, as measured by steering wheel, accelerator, and brake pedal usage, were also reported in the fast music condition relative to the no music baseline condition.

Other researchers have reported a modulation of mild driver aggression as a driver listens to their favourite music while driving. For instance, Wiesenthal, Hennessy, and Totten (2003) studied the potential distraction and relaxation effects resulting from listening to music on 40 volunteer drivers as they commuted from home to school or work. Under conditions of high traffic congestion, mild driver aggression (as measured by the State Driver Aggression Questionnaire; see Hennessy and Wiesenthal, 1999) was lowered when the drivers listened to their favourite music and when the time pressure was low, but not when the drivers were under high time pressure (or under conditions of low traffic congestion). Thus, it appears that the induction of relaxation in drivers by music may only be apparent under a subset of road conditions and when other external factors are not salient. Similarly, in an earlier study, Wiesenthal, Hennessy, and Totten (2000) also reported that listening to one's favourite music can result in lower driver stress than when driving in silence, at least under conditions of high congestion.

It is also worth noting at this point that the majority of the laboratory-based studies of the effect of music on driving have tended to test university students (Beh and Hirst, 1999; Brodsky, 2002; North and Hargreaves, 1999). It seems plausible that such younger drivers may be likely to be more influenced by the effects of background music than would older drivers. On the other hand, while listening to loud techno music has only been shown to lead to a mild decrement in the performance of younger drivers, one can only imagine the consequences for the older driver! Thus, further research is needed looking at the effects of different kinds of music for different age groups. Given that in real life, drivers are unlikely to spend too much of their time listening to music that they don't like, more studies should presumably also be conducted where drivers are allowed to listen to the music of their own choice (cf. Strayer and Johnston, 2001).

Given the marked differences in music preference (and familiarity; see Fontaine and Schwalm, 1979; Rentfrow and Gosling, 2003) between individuals, it is perhaps unsurprising that researchers have as yet been unable to draw any overarching conclusions regarding the specific effects of musical exposure on drivers. For instance, Furnham and Strbac (2002) reported differential distraction effects of background music and noise as a function of the particular personality traits of the drivers in their study (i.e., introverted vs. extraverted; see also Fagerstrom and Lisper, 1977; Rentfrow and Gosling, 2003). Meanwhile, Chamorro-Premuzic and Furnham (2007) conducted a questionnaire study on more than 340 individuals (university students) in order to look at individual differences and their relation to the importance of music in everyday life. Their results suggested that certain individuals, such as those who were neurotic and/or introverted were more likely to use music for emotional regulation (i.e., to change or enhance their mood). It is also worth noting that it has been reported that men and women are differentially affected by the emotional content of music, with women typically tending to be more responsive to the emotional content of music than men (e.g., Coffman, Gfeller, and Eckert, 1995; Kamenetsky, Hill, and Trehub, 1997; Panksepp, 1995).

One question here relates to whether a driver is given the choice to decide what music is played in the car, or whether instead he/she has to conform to a passenger's (or, for that matter, experimenter's) preferences (cf. Oldham, Cummings, Mischel, Schmidtke, and Zhou, 1995). In one laboratory-based study, Fontaine and Schwalm (1979) found that the more familiar a participant was with a particular piece of music, the bigger the beneficial effect of listening to that music on vigilance performance. Interestingly, however, the type of music itself had no significant effect on performance.

Over the years, researchers have investigated the effect of music on physiological (Hyde, 1924), cognitive (Boltz, Schulkind, and Kantra, 1991), and affective (Krumhansl, 2002) responses in a variety of different settings. Some researchers have even gone so far as to suggest that listening to certain kinds of music might make people temporarily more intelligent, an oft-cited effect that has come to be known as the 'Mozart effect' (cf. Schellenberg, 2001, for a critical review). Although Rauscher, Shaw, and Ky (1993) reported a short-term improvement in spatial-temporal reasoning ability after brief exposure to Mozart's Sonata for Two Pianos in D Major, K.448, the majority of subsequent research has failed to replicate the finding of any direct behavioural effect of this musical piece on spatial-temporal information-processing (see Schellenberg, 2001). Instead, the 'Mozart effect' appears to correlate primarily with changes in mood and/or arousal in participants as they listen to the specific piece of music. This may, as a consequence, influence people's performance in certain cognitive tasks, rather than this specific piece of Mozart's music generating any specific enhancement in spatiotemporal information-processing per se (see e.g., Ho et al., 2007a, on this point).

Taken together, the results of these studies do not seem to point to any specific performance deterioration that can be attributed to listening to music (cf. Beh and Hirst, 1999). Note though that music, or as a matter of fact, any other auditory stimulus that is played too loudly, may hinder performance, as suggested by the basic concept of the inverted U-shaped arousal function for performance (first proposed by Yerkes and Dodson, 1908; see Figure 3.1; see also Arent and Landers, 2003; Davenport, 1972).

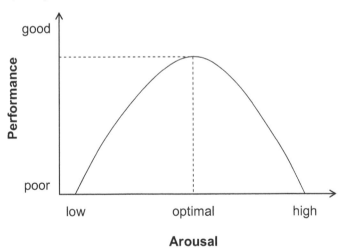

Figure 3.1 **Yerkes and Dodson's (1908) inverted U-shaped arousal function of performance states that performance on a given task improves with arousal until an optimal level, after which higher levels of arousal give rise to poorer performance**

It is not uncommon for researchers to assume low distraction caused by the radio primarily because so few of the available accident records suggest the use of the radio as a direct cause of car accidents. This may render a false impression when researchers attempt to compare the degree of distraction caused by mobile phone use to that of radio use. In addition, to date, the majority of studies appear to show limited interference from listening to the radio on driving performance (e.g., see Strayer and Johnston, 2001). For instance, Consiglio, Driscoll, Witte, and Berg (2003) reported slower braking reaction times in a laboratory-based task where their participants had to depress the brake pedal following a red brake light signal when they were conversing either with a passenger, or using a handheld or hands-free mobile phone. However, the same study failed to find any significant effect of listening to music on braking performance.

Shifting attention between different regions of space

More interestingly, certain researchers have even proposed that when drivers talk on the mobile phone while driving, they may have to shift their attention constantly between their personal space (the conversation) and the peripersonal and extrapersonal spaces (driving tasks; Fagioli and Ferlazzo, 2006; see Chapter 2). If this shift of attention to the personal space of the driver really does take place, then one might wonder whether the same thing might also apply to the case of a driver listening to the car radio. In which case, one might expect an unfavourable effect of listening to the radio on driving (see Chapter 8 for a further discussion of the distinct functional regions of space).

Driving with the radio on

Research over the last 40 years has identified a number of potential dangers and limitations associated with listening to the car radio, including increased risky driving behaviours (e.g., Brodsky, 2002). Several aspects of music itself have been shown to have an important (i.e., significant) effect on driver performance, including variations in the volume (e.g., Beh and Hirst, 1999), type, or tempo of the music (e.g., Brodsky, 2002), as well as the driver's liking for the music itself (e.g., North and Hargreaves, 1999), and above all, these factors modulate (and interact with) the mental state of the driver (such as whether they are drowsy vs. alert, and also their mood). Given that the majority of these studies were conducted in a laboratory setting, further investigations on the road or with the use of advanced driving simulators are needed in order to validate the potentially beneficial or adverse effects of music on driving (cf. Haigney and Westerman, 2001). As has been noted already, the majority of published studies used only young participants (mainly undergraduate students), so limiting the extent of the generalization of the findings thus far to the general driver population (see also Chamorro-Premuzic and Furnham, 2007). It will therefore be particularly important to examine how listening to the radio may impact on a driver's responses to auditory warning signals owing to its significant safety implications.

Chapter 4

The Auditory Spatial Cuing of Driver Attention

Introduction

Recent evidence from both behavioural and electrophysiological studies suggests that human performance on a variety of visual tasks can be facilitated by the presentation of an auditory cue from the same location as the visual stimuli shortly beforehand, even when the auditory cue does not provide any information about the visual task itself (see Spence et al., 2004, for a review). It thus seems plausible to extend these robust findings concerning the ability of spatial auditory cues to exogenously capture a person's visual attention in the laboratory setting to the design of multisensory interfaces, for instance, in the domain of developing effective warning signals for car drivers (e.g., Spence and Driver, 1999).

A few early interface design studies that were focused on trying to improve air cockpit or control room design investigated the potential performance advantages that might be offered by the presentation of spatially-localized auditory warning signals in an applied setting. For instance, in one oft-cited study by Begault (1993), air pilots were instructed to perform a visual search task for targets presented through the window of a cockpit with the aid of spatial auditory cues that were always presented from the same direction as the visual targets (that is, the spatial auditory cues were 100 per cent informative). In particular, the spoken air traffic alert 'traffic, traffic' was presented over a 3-dimensional head-up auditory warning display (as typically used by flight crews) from one of seven possible target locations. A significant improvement in visual search times of up to 2200 ms (without any loss of accuracy) was reported when the pilots performed the task using the spatial auditory display, as compared to control participants who performed the task with nonspatial auditory cues presented to only one ear (though see also van der Burg, Olivers, Bronkhorst, and Theeuwes, forthcoming). Not only does this large facilitation of visual search times have important implications in terms of safety for pilots who may be trying to avoid an approaching object, the large magnitude of the effect reported (as compared to that typically seen in laboratory studies; e.g., Spence, McDonald, and Driver, 2004) demonstrates that insights obtained from crossmodal spatial cuing studies conducted in the laboratory may prove to be of considerable value in real world settings.

Other studies by Perrott and his colleagues (e.g., Perrott, Saberi, Brown, and Strybel, 1990; Perrott, Sadralodabai, Saberi, and Strybel, 1991) have also demonstrated a marked facilitation of visual search performance when the participants in their study were aided by the presentation of a spatially correlated auditory cue (a 10 Hz click train). Similarly, 2- and 3-dimensional auditory spatial cues and 2-dimensional visual cues have been shown to be particularly effective in terms of their ability to enhance a person's ability to search for a visual target as compared to the baseline condition in which no cue was provided (Perrott, Cisneros, McKinley, and D'Angelo, 1996). Taken together, the results of these studies demonstrate that interface designers may potentially benefit from incorporating what psychologists and neuroscientists know about crossmodal links in spatial attention between audition and vision (and presumably other modalities, such as touch, as well; e.g., see Chapter 5) in the design of future applications (see also Spence and Driver, 1997a).

Auditory spatial cuing experiments

From an applied perspective, it is important to try and distinguish between the potential benefits offered by the exogenous orienting of spatial attention and those offered by the endogenous orienting of attention. The facilitatory effects reported in these earlier applied studies (e.g., Begault, 1993; Perrott et al., 1990, 1991, 1996) involved the use of spatially-predictive auditory warning signals. It is therefore unclear whether the performance advantages observed in these studies reflect the consequences of crossmodal links in exogenous orienting (i.e., driven by the physical positioning of the warning signal in space), the consequences of endogenous orienting (i.e., driven by the informative content of the warning signal), or the consequences of both types of orienting being engaged simultaneously by the warning signals. If any improvement in performance is principally driven by endogenous orienting, it would suggest that any symbolic warning signals that provide spatial content may give rise to performance benefits, and that the actual (or virtual) location from which the warning signals are presented may be rendered irrelevant. In this chapter, we will therefore illustrate how interface designers can better understand the potential benefits offered by the multisensory orienting of spatial attention with reference to a series of experiments from our own laboratory that were designed to examine the auditory spatial cuing of visual attention in a complex visual environment involving a simulated driving task (see Table 4.1). In particular, these experiments were designed to investigate whether the location and/ or informational content of an auditory warning signal could be used to facilitate the detection of potential emergency situations seen in the rearview mirror and/or from the front in a driving situation (Ho and Spence, 2005a).

Graham (1999) has suggested that the sound of a car horn does not necessarily have any inherent association with dangerous driving situations, which might, for example, be implicit in the sound of skidding tires. Instead, he suggested that the sound of a car horn is typically understood by drivers as indicating the presence of another vehicle in the vicinity (cf. Edworthy and Hellier, 2006). Experiment 4.1 was therefore conducted in order to assess the general alerting effect elicited by the presentation of a car horn sound for drivers. It should be pointed out that the majority of previous studies that have attempted to investigate the nature of any crossmodal links in spatial attention involving audition have typically tended to use particularly abstract sounds, such as, for example, white noise bursts or pure tones. The effectiveness of the nonspatial-nonpredictive (in terms of predicting the direction of the target event) auditory cue

Table 4.1   Summary list of the experimental conditions used in Experiments 4.1–4.5

Experiment	Cue sound	Cue type	Cue validity
4.1	Car horn	Nonspatial (alerting)	N/A
4.2	Car horn	Spatial	50%
4.3	Car horn	Spatial	80%
4.4	Verbal directional cues	Symbolic	80%
4.5	Verbal directional cues	Spatial symbolic	80%

presented in Experiment 4.1, which was always presented from the same spatial location directly under the participant's seat, in alerting/redirecting visual attention served as a baseline measurement for comparison with the results of the subsequent experiments in which spatial auditory cues were used. These auditory cues were 'spatial' in the sense that the cues coincided with, or predicted, the direction of the target visual driving events on a certain percentage of the trials. It was hypothesized that the spatially-neutral cue might act as a nonspatial alerting signal (see Posner, 1978; Spence and Driver, 1997b; Zeigler, Graham, and Hackley, 2001) that would result in a general facilitation of response latencies in the visually-mediated driving task.

Next, Experiment 4.2 investigated the effectiveness of the presentation of spatially-nonpredictive car horn sounds in orienting a person's attention to subsequent visual driving events presented either from in front or from behind them (and seen via a rearview mirror). In particular, the spatial auditory cue was spatially-nonpredictive in the sense that it was presented from the same direction as the target visual driving event on 50 per cent of the trials, and from the inappropriate (i.e., opposite) direction on the remainder of the trials. In effect, the spatial auditory cue did not provide any information about the nature of the task. The responses to target driving events occurring in the cued direction were expected to be faster than those taking place in the uncued direction, even though the cues were uninformative with regard to the likely location of the target visual events. We could then attribute any such effect that was obtained exclusively to the exogenous orienting of our participants' spatial attention in the direction of the auditory warning signals.

Experiment 4.3 further investigated whether the effectiveness of such spatial auditory cues would be enhanced if they were made predictive of the likely location of an upcoming visual driving event. The participants in our third experiment were informed that the direction from which the car horn sound would be presented would correctly predict the location of the target visual driving event on 80 per cent of the trials. Any increase in the magnitude of the interaction between the relative locations of the auditory and visual stimuli in Experiment 4.3 as compared to that of participants in Experiment 4.2 would then be attributable to the combined effect of the exogenous and endogenous orienting of spatial attention (see Spence and Driver, 1994, 2004).

Experiment 4.4 was designed to investigate the relative effectiveness of symbolic spatial verbal instructions (i.e., the words 'front' and 'back') in orienting a person's spatial attention when the cues were made 80 per cent predictive of the likely location of the target visual driving event. Taking the hypothesis one step further, Experiment 4.5 then went on to examine the potential efficacy of presenting redundant spatial cues that were composed of the symbolic spatial verbal instructions used in Experiment 4.4 presented from their respective spatial direction (i.e., the word 'front' presented from the front and the word 'back' presented from the back).

We simulated dual-task driving conditions (see Figures 4.1 and 4.2). Specifically, participants in the experiments reported in this chapter had to perform a rapid serial visual presentation (RSVP) task designed to simulate a continuously and uniformly

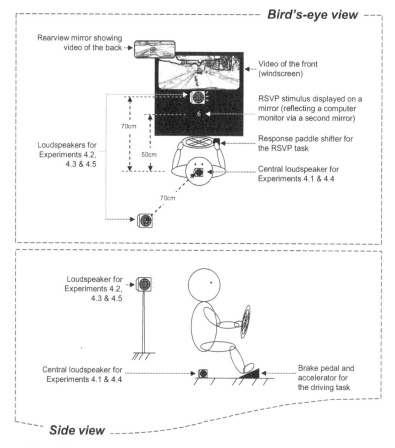

Figure 4.1   Schematic diagram of the experimental set-up used in Experiments 4.1–4.5

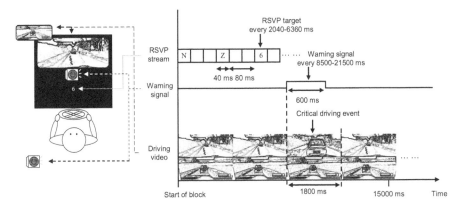

Figure 4.2   Schematic timeline showing the temporal sequence of events in Experiments 4.1–4.5

highly attention-demanding situation (see Shapiro, 2001). The RSVP task used in the experiments reported in this chapter consisted of the presentation of a continuous stream of distractor letters (B, C, D, E, F, J, K, L, M, N, P, R, S, T, X, Y and Z) with target digits (taken from the set: 2, 3, 4, 5, 6 and 9) periodically embedded within the stream. A total of 66 target digits were presented in each of the eight blocks of experimental trials. The temporal gap between successive target digits was in the range of 2040 – 6360 ms.

At the same time, the participants in our experiments had to respond to randomly-presented visual driving events in a simulated driving task. The video clips depicting the actual views of the windscreen and rearview of a driven car in a car following scenario were presented to the participants via two video monitors. The critical visual driving events consisted of the sudden rapid approach by the 'driven' car toward the car in front at approximately 100 km/h, or the sudden rapid approach by the car behind. The rapid approach toward the lead car was presented via the video monitor positioned directly in front of the participants, while the rapid approach of the trailing vehicle was presented on the video monitor positioned behind the participants and seen by the participants via looking in the rearview mirror. The participants had to respond to the 24 visual driving events per block with the aid of auditory cues which consisted of the sound of a real car horn being presented for 600 ms. The temporal gap between successive auditory cues was in the range of 8500–21500 ms. The participants gave their responses to the RSVP task by moving the right paddle shifter on the steering wheel toward themselves. They responded to the driving task by pressing down on the accelerator (following the approach of the trailing vehicle) and depressing the brake pedal (following their rapid approach toward the lead vehicle).

In real-world driving, a driver's visual attention is primarily focused toward the front, with events at the rear typically only monitored by means of the rearview and side mirrors (cf. Binkofski, Buccino, Dohle, Seitz, and Freund, 1999; Burnett and Joyner, 1997). Even so, there are blind spots around a vehicle that will fall outside a driver's line of sight at any given moment. The attention paid to the rear therefore depends, to a great extent, on the frequency with which a driver checks his/her mirrors. In one of the few studies to have been conducted in this area, Brookhuis, de Vries, and de Waard (1991) evaluated the effects of introducing dual-task demands (in this case, the use of a mobile telephone) on driving performance (see also Chapter 2). They used rearview mirror-checking as a performance measure to estimate how much attention participants paid to the other traffic. The researchers found that the frequency of mirror checking depended primarily upon the road situation (e.g., on a busy ring-road, less attention was paid to the mirrors as compared to the situation in which the person was driving on a quiet motorway). Interestingly, the dual task demands of telephoning and driving simultaneously did not seem to deleteriously affect the frequency of mirror checking (i.e., of attention being directed toward the rear).

Given that most drivers have extensive experience of the fact that the rearview mirror represents the space behind their car, an auditory signal from the rear might be expected to facilitate the detection of a critical situation from behind albeit seen from the front via the rearview mirror. Such an hypothesis is based on the assumptions that: 1) crossmodal links in exogenous spatial attention between audition and vision can facilitate the appropriate deployment of a person's spatial attention; and 2) there appears to be an automatic and effortless translation between visual events seen in front (but via the rearview mirror) and the awareness of driving events occurring at the rear.

Tables 4.2 and 4.3 provide a summary of the results of the auditory warning signal experiments that have been conducted in our laboratory in Oxford. The

Table 4.2   Mean reaction times (RTs; ms) and percentages of errors in the driving task reported as a function of the location of the visual driving event in Experiments 4.1–4.5

				Front (windscreen)		Back (rearview mirror)	
Experiment	Cue sound	Cue type	Cue validity	RT (ms)	% errors	RT (ms)	% errors
4.1	Car horn	Nonspatial (alerting)	N/A	911	4.9	1083	5.1
4.2	Car horn	Spatial	50%	1115	5.8	1133	5.9
4.3	Car horn	Spatial	80%	970	4.0	1046	3.7
4.4	Verbal directional cues	Symbolic	80%	805	9.6	985	7.6
4.5	Verbal directional cues	Spatial symbolic	80%	780	10.7	957	8.5

Table 4.3 **Mean reaction times (RTs; ms) and percentages of errors in the driving task as a function of the location of the visual driving event and the direction of the auditory cue in Experiments 4.2–4.5**

				Location of the visual driving event							
				Front (windscreen)				Back (rearview mirror)			
				Front auditory cue		Rear auditory cue		Front auditory cue		Rear auditory cue	
Experiment	Cue sound	Cue type	Cue validity	RT	% error	RT	% error	RT	% error	RT	% error
4.2	Car horn	Spatial	50%	1106	6.3	1124	5.2	1168	5.5	1098	6.2
4.3	Car horn	Spatial	80%	907	1.8	1033	6.3	1107	3.7	985	3.7
4.4	Verbal directional cues	Symbolic	80%	770	2.7	840	16.5	1025	10.9	945	4.3
4.5	Verbal directional cues	Spatial symbolic	80%	695	2.0	865	19.5	1052	13.0	862	4.1

results of Experiment 4.2 reveal a significant performance advantage when the participants responded to critical visual driving events seen in the rearview mirror that were preceded by an auditory cue from the rear (i.e., from the same direction) rather than from in front (i.e., the opposite direction; see Figure 4.3A). A spatial cuing effect was observed following the presentation of spatially-nonpredictive car horn sound cues in Experiment 4.2. At one level, these spatially-nonpredictive car horn sounds can be considered as being very similar to the nonspatial-nonpredictive sounds used in Experiment 4.1 (cf. van der Burg et al., forthcoming) since both were spatially uninformative with regard to the likely location of the target visual driving events. However, the redundant information elicited by the spatial variability in the location of the car horn sound might have required extra time to process, thus potentially slowing down participants' reactions (see Wallace and Fisher, 1998). This suggests that the invalid or incongruent spatial directional cue can impair performance while the correct spatial directional cue may facilitate it.

It should, however, be noted that the driving task used in the experiments reported in this chapter was a relatively simple one. That is, the participants only had to judge whether there would be a potential collision (to the front or rear) and react appropriately to this information by accelerating, braking, or else by making no response. This may have limited the magnitude of the facilitatory effect of the spatial cues because the participants only had to distinguish between four possible road scenarios (i.e., front-critical, rear-critical, front-noncritical and rear-noncritical), which may have been quite easy for them to learn. In a real driving situation, by contrast, drivers would presumably need more time in which to perceive and analyse the situation and so give an appropriate response even if a spatial warning signal can direct their visual attention in a particular

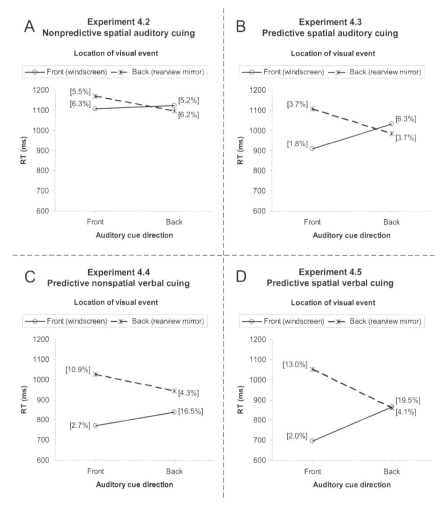

Figure 4.3   Summary interactions in the mean reaction time (RT; ms) data between the direction of the auditory cue and the location of the critical visual event in the driving task in the auditory spatial cuing Experiments 4.2 (A), 4.3 (B), 4.4 (C) and 4.5 (D). Percentages of errors are indicated by numerical values in parentheses. The slope of the lines indicates the magnitude of the spatial cuing effects that were observed

(presumably appropriate) direction. It has been argued that crossmodal links in spatial attention may become more pronounced in general as task load increases (e.g., see Spence, Ranson, and Driver, 2000b). Note also that when drivers are on the road, they typically have to monitor the information available to many different senses simultaneously (e.g., not just vision and audition, as studied here, but also proprioception and vestibular inputs; see Kemeny and Panerai, 2003), and hence the attentional load of actual driving may be higher (e.g., see Lavie, 2005; Rees and Lavie, 2001; though see also Sivak, 1996).

The spatial auditory cues presented in Experiment 4.2 may be criticized in terms of a realistic driving situation for their lack of reliability (being only 50 per cent valid in terms of informing the participants of the likely location of the target event). Since unreliable cues have been shown to adversely affect performance in real life situations (e.g., see Bliss and Acton, 2003; Sorkin, 1988), Experiment 4.3 was designed to examine the effectiveness of informative spatial cues (i.e., cues that were 80 per cent predictive with regard to the likely location of the target visual driving event) to model ecologically-valid situations such as in real life when warning cues are designed to be informative (cf. Sorkin, 1988).

The results of Experiment 4.3 support the potential usefulness of spatially informative auditory cues in terms of their ability to capture a driver's visual attention (see Figure 4.3B). In particular, the participants responded significantly more rapidly to target visual driving events presented both from the front and from the rear when they were preceded by a valid than by a spatially invalid spatial car horn sound.

In an early study of the relative effectiveness of various different auditory warning signals, Oyer and Hardick (1963) investigated the optimum acoustical dimensions of auditory alerting signals. Sounds including animal and human noises (such as elephants stampeding and the sound of a baby crying), sirens, bells, horns, buzzers, whistles, pure tones and noises of various types were assessed for their potential alerting effect by different segments of the population of the United States. The participants had to rate the sounds along the dimensions of pleasant-unpleasant, startling-nonstartling and bizarre-ordinary. On the basis of their results, Oyer and Hardick were able to specify the frequency, intensity, and time ranges within which the optimal warning signals should appear. From amongst all of the many different alerting sounds that were evaluated in their large-scale study, one of the car horn sounds was rated as being the fifth most effective alerting signal, preceded only by the sounds of a missile alarm, yelper siren, British air raid siren and falcon horn.

Note though that the current approach to ergonomic alarm design has tended to put a greater emphasis on the informative nature of alarms (as opposed to their capacity simply to alert as was the case back in the 1960s; cf. Edworthy and Hellier, 2006). The spatial auditory warning signals used in Experiments 4.2 and 4.3 lie closer to the informational end on the inform-alert continuum, as the cues were not designed simply to startle the participants, but to provide spatial information with regard to the visual target events (though this was less true in Experiment 4.2). Interface designers need to examine the effectiveness of different warning signals further when specifying the most appropriate signals for the particular applications concerned. This needs to be done both in terms of their ability to alert or arouse an interface operator, and also in terms of their ability to direct people's attention to the locations of interest both internally (e.g., on the interface display itself) and externally (e.g., by alerting a driver to any potential road danger).

Given that words are used to convey explicit meanings in everyday language, Experiment 4.4 was designed to investigate whether the verbal cues (the words 'front' and 'back') could be used to examine the effectiveness of verbal instructions in orienting a driver's visual attention (see Figure 4.3C). The results of this experiment suggested that participants could rapidly comprehend the meanings of the verbal warning signals and direct their spatial attention appropriately. Verbal warning signals have been shown to be more effective than non-speech abstract warnings in that they carry more information about the upcoming event, in particular in situations when the operator is under a high cognitive load (e.g., Graham, 1999; Selcon et al., 1995). With reference to Figure 4.3 (compare panels B and C), it can be seen that when cued by a verbal instruction to the likely position of the visual event, the participants responded more rapidly than when nonverbal cues were used across all four combinations of auditory cue direction and visual stimulus location. This was, however, accompanied by an increase in errors as well. Taken together, and when considered in relation to Hirst and Graham's (1997) previous finding that speech warnings resulted in greater annoyance to the participants in their study, it would seem that the spatial and verbal predictive cues may have affected a driver's attention in different ways. The participants were faster, but less accurate, when responding following verbal cuing than when following nonverbal cuing.

Experiment 4.5 was designed to investigate whether combined verbal spatial warning signals (such as the word 'front' presented from the front and the word 'back' presented from the back) would perhaps be even more effective in orienting a driver's attention than the spatial or verbal cues used in Experiments 4.3 and 4.4 (cf. Selcon et al., 1995). The results show faster performance overall when the auditory cues were spatial-verbal (Experiment 4.5; see Figure 4.3D) than when they were nonspatial-verbal (Experiment 4.4) or else consisted of the spatial-nonverbal car horn sound cues (Experiment 4.3).

Taken together, the results of Experiments 4.1–4.5 highlight a number of the potential implications of the existence of crossmodal links in spatial attention between audition and vision for interface design. The results documented so far reveal that a spatially-nonpredictive auditory cue that is presented from an appropriate (or valid) spatial direction can be used to facilitate the overt orienting of a driver's visual attention in the same direction. The results of Experiments 4.2–4.5 therefore demonstrate a significant performance advantage in the detection of potential emergency driving events (requiring either an acceleration or braking response) when the participants were cued by an auditory cue coming from the relevant direction, or else when their attention was directed in the appropriate direction by means of a semantically-meaningful verbal cue. The beneficial effect of the presentation of the valid auditory spatial cue on our participants' performance was enhanced still further by making the cues spatially-predictive (as shown by comparing the data from Experiments 4.2 and 4.3). These results are consistent with Begault's (1993) findings, discussed earlier, that spatial auditory cues can be used to facilitate the visual search performance of pilots, and extend the potential facilitatory effect of the use of spatial auditory warning signals on visual information detection (and subsequent reactions) into a simulated driving set-up.

Overall, the results from the five experiments reported so far suggest that the combined use of exogenous and endogenous orienting offers the most effective means of capturing a driver's spatial attention. On the theoretical side, the demonstration of crossmodal links in spatial attention between audition and vision, and the varying effects of the various auditory cues on performance on the same visual task, together argue against traditional modality-specific accounts of independent channels for auditory and visual information processing (see Wickens, 1980, 1992, 2002; see also Chapters 1 and 2). Instead, the results reported in this chapter support recent claims that the mechanisms controlling spatial attention in different sensory modalities can interact and thereby influence one another (e.g., Spence and Driver, 2004; Spence and Gallace, 2007). Given that the facilitation of responses was measured at the behavioural level, it will be an important question for future research to investigate whether the performance improvement reported in Experiments 4.2–4.5 reflects the consequences of a perceptual enhancement attributable to the spatial aspect of the cues coinciding with that of the targets, a priming of the appropriate responses by the cues (which can be generated by cues regardless of their location relative to the critical driving events) or some unknown combination of these two effects (see Chapter 6 for further experiments on this topic). No matter what the underlying cause(s) of the facilitatory effects reported in this chapter turn(s) out to be, it is important for the applied domain to combine the perceptual and response compatibility aspects of cuing in the design of the most effective multisensory interfaces (and warning signals).

Meaningful auditory warnings

Previous studies that have attempted to investigate the nature of the crossmodal links underlying exogenous spatial attention in humans have always used cues of questionable (or no) ecological validity, such as pure tones or white noise bursts in the case of auditory cuing experiments (see Spence et al., 2004, for a review). It is therefore possible that the use of a semantically-meaningful sound, such as the car horn used in Experiments 4.1–4.3, might engage a person's attention more effectively than these other arbitrary sounds (cf. Oyer and Hardick, 1963). It remains an interesting question for future research to determine whether the use of a car horn or other semantically-meaningful spatial auditory cue will prove even more effective at capturing an interface operator's attention than, say, a white noise burst or pure tone of equivalent intensity and localizability. Note that the semantic meaning conveyed by the car horn sound used in Experiments 4.1–4.3 might also have activated other attentional mechanisms than those typically thought to control spatial attentional orienting (cf. Langton, Watt, and Bruce, 2000). For instance, research has shown that the physiological state of arousal of drivers can sometimes determine their driving behaviours, for instance by modifying their responses to critical emergency driving situations (see Collet, Petit, Priez, and Dittmar, 2005; cf. Johnston and Shapiro, 1989).

In his extended analysis of mirrors and perception, Gregory (1998) mentioned the visual ambiguity in certain images perceived via mirror reflection. In particular, he hints that in an emergency situation, drivers may need to spend time processing what they perceive in the rearview or side mirrors (see Gregory, pp. 208–9). In fact, if they are lucky, drivers may not encounter a potential emergency situation for several years, but when they do, the warning signal needs to alert the driver and convey sufficient information for him/her to make the necessary reaction in a cognitively demanding and short time frame (cf. Belz, Robinson, and Casali, 1999).

False alarms

The issue of warning signal reliability has been investigated in both the aviation and driving literature (e.g., Bliss, 2003; Parasuraman et al., 1997; Xu, Wickens, and Rantanen, 2007). One key point to have emerged from this research is that false alarms can potentially be very distracting to an operator in diverting attention from the primary task. Additionally, the available evidence suggests that an operator may choose to ignore a warning signal if it proves to be inconsistent and/or unreliable, thus potentially negating its value (e.g., Parasuraman and Riley, 1997; Sorkin, 1988).

For example, in one recent study, Bliss and Acton (2003) examined the effect of varying the reliability of collision alarms on performance in a driving simulator. The participants in this study were instructed to drive and to swerve to the left or right depending on the side from which cars approached randomly in the rearview mirror. Bliss and Acton reported that performance improved (i.e., in terms of the frequency of appropriate reactions by the drivers to the alarm signals) as the reliability of the cue increased (from 50 per cent through 75 per cent to 100 per cent validity). Consistent with Bliss and Acton's findings, responses were both faster and more accurate when the reliability of the auditory cues increased from 50 per cent valid (Experiment 4.2) to 80 per cent valid (Experiment 4.3) in the present study.

Speech warning signals

A person's response to a speech warning may also be slower in an emergency situation than under normal circumstances, particularly if an operator does not understand the meaning of a speech message until it has finished (e.g., Simpson and Marchionda-Frost, 1984). Since Experiments 4.4 and 4.5 used only two single-syllable verbal cues ('front' and 'back'), the results were probably not influenced by any potential problems associated with disambiguating multiple possible speech warnings that may be present in applied settings that are likely to incorporate multiple verbal warning signals (cf. Chan, Merrifield, and Spence, 2005; Ho and Spence, 2006).

Given that drivers are typically engaged in other tasks involving linguistic elements, such as, for example, listening to the radio (see Chapter 3), perhaps having a conversation with a passenger or over the mobile phone (see Chapter 2), or else possibly taking in speech instructions from an in-car navigation system (e.g., Dingus et al., 1997; Kames, 1978; Streeter, Vitello, and Wonsiewicz, 1985), the additional verbal cues from the warning systems may inevitably lead to some confusion, misunderstanding and/or masking (either energetic or informational; see Chan et al., 2005; Oh and Lutfi, 1999). Thus, although drivers can easily acquire and react to a verbal instruction when it happens to be presented in isolation (as in Experiments

4.4 and 4.5), some caution may be needed before implementing verbal cues in a real warning system. In a real driving situation, it might actually be better to implement nonverbal warning signals, as they are less susceptible to the influence of other concurrent linguistic elements that may be present in the environment.

Some recent research was specifically designed to investigate the most effective audible icons that can convey the 'right degree of urgency' and have a 'commonly-understood meaning' for use in in-car systems (e.g., Catchpole, McKeown, and Withington, 1999a, b; McKeown and Isherwood, 2007). Indeed, there is a long history of research into the idea of using ecologically-valid auditory icons in interface design (see Catchpole et al., 2004; Gaver, 1993a, b; Oyer and Hardick, 1963), although that has not as yet been widely implemented in in-car systems. Spatial cues may therefore provide a more promising line for inquiry than verbal cues or arbitrary nonspatial auditory icons in that they utilize an orienting response that should be common to everyone, whereas there may be distinct cultural interpretations of the meanings inherent in different auditory icons. A successful warning signal needs to be representational in the sense that any likely operator can intuitively recognize its meaning and be informed of the required actions.

Many people play their music at levels that are simply too loud (i.e., more than 83 dB) while driving (Ramsey and Simmons, 1993; see Chapter 3). For not only can loud music mask other driving noises in the environment, such as symptomatic engine sounds and warning signals that are important for driving (Booher, 1978; Ho and Spence, 2005a; though see also McLane and Wierwille, 1975); there is also a danger that future auditory warnings will need to be presented at even higher levels in order to compensate for the growing population of drivers suffering from hearing loss (cf. Tannen, Nelson, Bolia, Warm, and Dember, 2004, on the notion of adaptive displays; see also Slawinski and MacNeil, 2002). Problems associated with playing music in the car are also expected to increase given the recent increase in the use of auditory warning signals designed to overcome visual overload (Belz et al., 1999; Ho, Reed, and Spence, 2007; Ho and Spence, 2005a), as music may potentially obscure the audibility of auditory warning signals. It thus seems essential that researchers study the possibility of utilizing the driver's other senses, such as their sense of touch (e.g., through the increased use of vibrotactile warning signals), when driving. This is where we will turn in the next chapter.

Chapter 5

The Vibrotactile Spatial Cuing of Driver Attention

Introduction

The human skin constitutes about 18 per cent of our body mass (Montagu, 1971), yet the importance of our sense of touch in driving has frequently been overlooked in the past (see Gallace et al., 2007, for a recent review of the literature on touch). Recently, however, there has been a rapid growth of interest, both theoretical and commercial, in the potential use of tactile interfaces in a variety of applied environments (e.g., Gilliland and Schlegel, 1994; Jones et al., forthcoming; Rupert, 2000a; Sklar and Sarter, 1999; Spence and Ho, submitted; Triggs, Lewison, and Sanneman, 1974; van Erp et al., 2004; van Erp and van Veen, 2006; Zlotnik, 1988). This recent growth of interest has been driven, in part, by the increasing visual overload reported by interface users (and car drivers; see Figure 5.1; e.g., Senders, Kristofferson, Levison, Dietrich, and Ward, 1967; Sivak, 1996; Sorkin, 1987; Zlotnik, 1988). As a result, many researchers have started to investigate the implementation of tactile devices into contemporary interface design in an attempt to share the workload exerted on our visual systems. A number of different tactile applications have been tested in a wide range of user environments such as, for example: tactile-visual sensory substitution (TVSS) systems to provide information to assist visually-impaired individuals (e.g., Bach-y-Rita, 2004; White, 1970); navigational aids to provide directional or way-finding information to pilots and drivers (e.g., Eves and Novak, 1998; Kuc, 1989; Tan, Gray, Young, and Traylor, 2003; van Erp et al., 2004; van Erp and van Veen, 2001; van Veen, Spape, and van Erp, 2004); tactile orientation awareness displays for use in altered sensory environments such as in microgravity environments or while deep-sea diving (e.g., Bhargava et al., 2005; Rochlis and Newman, 2000; Rupert, 2000a; Traylor and Tan, 2002; van Erp and van Veen, 2003); and haptic virtual reality displays that can, in some sense, simulate the sensation of touch (Wood, 1998).

Psychophysical research on touch dating back to the 1960s has studied variations in perceptual thresholds for tactile stimuli delivered to different body sites (e.g., see Deatherage, 1972; Geldard, 1960; von Békésy, 1963; Weinstein, 1968). Meanwhile, other studies have investigated people's detection sensitivity and their ability to localize tactile stimuli, with the majority of studies presenting tactile stimuli on people's forearms, hands and/or their fingertips (e.g., Burke et al., 1980; Jagacinski et al., 1979; Schumann et al., 1993; Sklar and Sarter, 1999; Vitense et al., 2003) and also around the waist in certain more recent application-inspired research (e.g., Cholewiak, Brill, and Schwab, 2004; Ho and Spence, 2007; Nagel, Carl, Kringe,

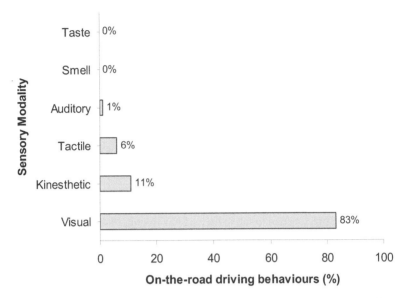

Figure 5.1   Sivak (1996) evaluated 89 of the most critical on-the-road behaviours in terms of the sensory modality that was required for their occurrence. Interestingly, 27 of these behaviours depended on input from more than one sense. (Note that the kinaesthetic category presumably includes behaviours that rely on proprioceptive and/or vestibular inputs as well)

Martin, and Konig, 2005; van Erp and van Veen, 2004). As for in-car tactile interfaces, recent studies have begun to investigate the possibility of stimulating the driver's body via tactile displays embedded in the driver's seat (e.g., Fitch et al., 2007; Van Erp and van Veen, 2004), worn around their waist (e.g., Ho, Reed, and Spence, 2006a, 2007b; Ho, Tan, and Spence, 2005b), via the steering wheel to the driver's hands (e.g., Enriquez, Afonin, Yager, and Maclean, 2001; Steele and Gillespie, 2001) and via the accelerator or brake pedal to a driver's feet (e.g., Godthelp and Schumann, 1993; Janssen and Nilsson, 1993; Janssen and Thomas, 1997; Kume, Shirai, Tsuda, and Hatada, 1998). Other relevant examples here include a shape-changing haptic controller (Michelitsch, Williams, Osen, Jimenez, and Rapp, 2004) and a haptic steering wheel that has been designed to assist people when driving around corners and overtaking (Schumann, Godthelp, and Hoekstra, 1992; Schumann and Naab, 1992).

Table 5.1 highlights some of the potential advantages associated with the use of tactile communication over visual and auditory communication as reported in the literature. Given these potential advantages, we decided to conduct a series of tactile cuing experiments in an attempt to investigate the possibility of using spatially-distributed vibrotactile warning signals in order to direct a person's attention to the front versus rear in a simulated car driving task (Ho et al., 2005b; cf. Enriquez and MacLean, 2004; Fenton, 1966; van Erp and van Veen, 2001). Once again, we chose the front-to-rear-end collision scenario, because this kind of accident is particularly

Table 5.1   Potential advantages of tactile communication highlighted in the literature

Purported advantages of tactile communication	Study
Rapid transmission of tactile information to the brain.	Harrar and Harris (2005); Mowbray and Gebhard (1961)
Relatively automatic ability to alert.	Geldard (1960); Gilmer (1961)
Inherent directionality that is mapped naturally to bodily coordinates.	van Erp and van Veen (2001)
Unlike vision, tactile (and auditory) system cannot be switched off voluntarily.	Gilmer (1960, 1961)

common amongst drivers who are distracted by their in-car technology (such as by, for example, the mobile phone, McEvoy et al., 2007a; Strayer and Drews, 2004; see Chapter 2). This series of experiments was motivated by recent laboratory-based studies that have demonstrated the existence of robust crossmodal links in spatial attention between vision and touch (see Chapter 1; Spence and Driver, 2004, Spence and Gallace, 2007, for reviews). For example, studies have shown that the presentation of a vibrotactile cue to one side or the other results in a short-lasting shift of attention to the cued direction, and facilitates the processing of stimuli subsequently presented from the cued direction (e.g., Kennett, Eimer, Spence, and Driver, 2001; Kennett, Spence, and Driver, 2002; Spence, Nicholls, Gillespie, and Driver, 1998; Tassinari and Campara, 1996). However, the magnitude of the observed cuing effects has typically been quite small (e.g., from a maximum cuing effect of 10 ms in Tassinari and Campara's 1996 study to 48 ms in Spence, Pavani, and Driver's 2000a studies). Thus, it is uncertain as to whether these theoretically-motivated laboratory-based studies necessarily provide any results that should merit the interest of those in the applied community. When these studies are scaled-up from minimalistic laboratory environments (typically involving participants performing in dark and silent texting cubicles) and tested in more applied settings, such as in the case of multisensory warning signals for driving, the results can go in one of two directions. That is, the very small effects may either get lost in the presence of the noise that testing in a more ecologically valid situation can bring, or else the effects may get much larger as the prioritization of one position over another has a bigger impact (i.e., in environments where there are many different stimuli competing for a driver's attention, then helping to prioritize one stimulus over the others for preferential processing can lead to a big time saving).

Vibrotactile spatial cuing experiments

Two laboratory-based experiments (Ho et al., 2005b) were thus designed to assess the possibility of using spatially-distributed vibrotactile warning signals to direct a person's attention to the front or rear. Vibrotactile stimuli were presented via a belt worn around a participant's waist. In the first experiment, we examined performance under conditions where these vibrotactile stimuli were predictive (i.e., 80 per cent valid) with regard to the direction of a potential critical visual driving event, while the second experiment investigated the potential utility of nonpredictive vibrotactile cues (i.e., 50 per cent valid). We also conducted a driving simulator study (Ho et al., 2006a) in order to examine the potential use of vibrotactile cuing in driving in a more realistic testing environment.

Sixteen participants took part in our first vibrotactile cuing experiment. The simulated driving setup that we used was identical to that described previously in Chapter 4, with the exception that the participants now wore a vibrotactile belt around their waists (see Figure 5.2). One of the tactors was positioned in the middle of the participant's stomach, the other in the middle at their back. A vibrotactile warning signal, consisting of a 1060 ms vibration, was presented from either one of the two tactors at the onset of a visual driving event. The vibrotactile cues in this experiment correctly predicted the direction of the visual driving event on 80 per cent of the trials.

The results of Experiment 5.1 revealed that the participants responded both more rapidly, and somewhat more accurately, to critical visual driving events preceded by a valid vibrotactile cue presented from the same direction, as compared to those trials preceded by an invalid vibrotactile cue coming from the opposite direction (see Figure 5.3A). These results therefore suggest that the presentation of spatially-predictive vibrotactile cues in Experiment 5.1 led to a rapid crossmodal shift of visual attention in the direction indicated by the cue. In particular, vibrotactile stimuli presented on the body surface can be used to direct visual attention to distal events happening several metres away in extrapersonal space. Given the spatially-predictive nature of the vibrotactile cues used, it is possible that the participants might have used the informational content of the vibrotactile cues in order to direct

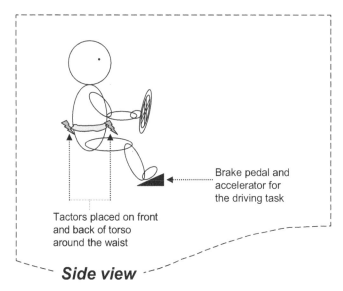

Figure 5.2   Schematic diagram of the experimental set-up used in Experiments 5.1 and 5.2 (see also Figure 4.1)

their attention endogenously in a top-down manner (rather than exogenously, or in a stimulus-driven manner). This implies that the informational content of the vibrotactile cues might be more important than their spatial location per se (see also Klein and Shore, 2000).

As a result, our second vibrotactile cuing experiment was designed to investigate whether vibrotactile spatial cuing remains effective when the vibrotactile cues were made nonpredictive, that is when they were just as likely to be presented from the same direction as the target critical visual driving events as from the opposite direction (i.e., when cue validity was 50 per cent). If the magnitude of the spatial cuing effects reported remained unchanged, then they could be wholly attributed to the exogenous attention-capturing effect of vibrotactile cues. If, however, the magnitude of the spatial cuing effect was substantially reduced, then this would suggest that the cuing effects reported in Experiment 5.1 should be attributable to the informational content of the cues that consequently gave rise to the endogenous orienting of spatial attention. Sixteen new participants took part in this experiment.

The results showed that the participants in Experiment 5.2 responded more rapidly and accurately to the critical visual driving events preceded by vibrotactile cues from the same spatial direction rather than from the opposite direction (see Figure 5.3B). Given that these cuing effects were elicited by cues that were spatially-nonpredictive with regard to the location of the critical visual driving events, this suggests that the presentation of the vibrotactile cues in Experiment 5.2 resulted in an exogenous shift of visual attention in the cued direction.

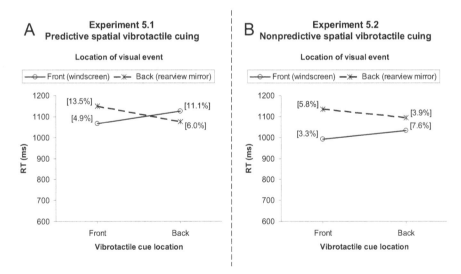

Figure 5.3 **Summary interactions in the mean reaction time (RT; ms) data between the direction of the vibrotactile cue and the location of the critical visual event in the driving task in the vibrotactile spatial cuing Experiments 5.1 (A) and 5.2 (B). Percentages of errors are indicated by numerical values in parentheses. The slope of the lines indicates the magnitude of the spatial cuing effects that were observed**

Once again, just as in the case of the auditory cuing experiments reported in Chapter 4, the facilitation of responses reported in this chapter was measured behaviourally. It will be important therefore from a theoretical perspective to investigate whether this performance enhancement reflects a consequence of response priming (i.e., a priming of the appropriate responses by the cues regardless of their spatial location relative to the critical driving events), a perceptual enhancement attributable to the spatial aspect of the cues coinciding with that of the targets, or some unknown combination of the two effects (see the experiments reported in Chapter 6 on this very issue). In applied terms, both aspects of performance enhancement are certainly important in terms of trying to design the most effective multisensory interfaces. Nevertheless, an understanding of the relative contributions of each of the two effects to performance should help to improve the efficacy of interface development in the future.

Taken together, the results of Experiments 5.1 and 5.2 converge with the findings of Sklar and Sarter's (1999) previous study of pilots in showing the potential utility of vibrotactile cues as warning signals in applied interface design. These results also highlight the safety implication of vibrotactile warning signals in automobile interface design, such as the design of more complex tactile warning systems for indicating one or more impending collisions (which may be particularly important for specialized industrial or agricultural vehicles; cf. McGehee and Raby, 2003; see also van Erp and van Veen, 2003; van Veen and van Erp, 2001).

One can interpret the driving task outlined in this and the preceding chapter as being similar to a car-following scenario on a highway where a driver is constantly attempting to maintain a safe distance both to the car in front and from the car behind, albeit with the experimental driving task constituting a simplified version of what drivers have to do in a real driving situation. The implementation of vibrotactile warning systems similar to the one outlined here for distance-keeping and/or lane-keeping may be particularly useful in situations such as on the highway when drivers may not pay attention to their increasing speed relative to the safety distance that they need to maintain in order to have sufficient headway should unexpected events occur in front (cf. Smith, Najm, and Lam, 2003).

It should, however, be borne in mind that in order to collect sufficient empirical data for statistical analysis in our experimental research, we had to present the warning signals (and the critical driving events that these warning signals were designed to inform the drivers about) far more frequently than they would be likely to be presented in any real-life setting. Several researchers have previously discussed the difficulty of capturing the 'completely unexpected' event that frequently leads to an accident in an experimental setting (no matter whether that be in the laboratory, as studied here, or in the simulator or on-road setting; see Haigney and Westerman, 2001; Tufano, 1997). Future research in this area will therefore need to ensure that the spatial cuing benefits that have been demonstrated in the experiments reported so far in this volume, still occur when the warning signals are presented far more infrequently (i.e., when their incidence of occurrence more closely matches realistic levels).

Bearing this potential caveat in mind, the results of the experiments reported in this chapter nevertheless suggest that the crossmodal links in spatial attention between touch and vision identified in previous laboratory-based studies may have an important application in a number of real-world cases in the future design of multisensory interfaces. Converging evidence from studies on pilots and astronauts has also demonstrated the effectiveness of tactile cues for spatial orientation, navigation and situational awareness (e.g., Gilson and Fenton, 1974; Rochlis and Newman, 2000; Rupert, 2000b; Traylor and Tan, 2002; Triggs et al., 1974). However, more research on the specific limitations of tactile communication is certainly needed before its widespread application in those areas where either no training or else minimal training is offered to users who need to operate an interface (hence the importance of the notion of 'intuitiveness' to warning signal design; see Rupert, 2000b; Spence and Gallace, 2007; Spence and Ho, submitted).

Ecological validity

In the laboratory-based computer driving simulation experiments reported in this and the preceding chapter, the RSVP task was used to simulate a continuously and uniformly highly attention-demanding situation (see Shapiro, 2001), such as when a driver's visual attention is concentrated entirely toward the front. This task has been used extensively in earlier dual-task attention research to maintain a high, and relatively constant, level of cognitive load on participants, and it has allowed researchers to measure the temporal distribution of attention under conditions of task-switching (e.g., Allport and Hsieh, 2001; Klein and Dick, 2002). However, in theory, this task might not necessarily have generated the same kinds of attentional demands as are involved in actual driving. Moreover, the driving stimulation was restricted to the visual, auditory and vibrotactile stimuli provided in the experiments, which might not be comparable to a realistic driving environment. It therefore seemed important to examine if the effects found previously could be reliably obtained under more realistic testing conditions, such as those offered by high-fidelity driving simulators (cf. Haigney and Westerman, 2001).

As such, our next experiment (Experiment 5.3) was specifically designed to investigate the possibility that driver responses to potential front-to-rear-end collision situations could be facilitated by implementing vibrotactile warning signals that indicate the likely direction of the potential collision in a driving simulator setting. If the results of Experiment 5.3 replicated those found in the two previous experiments, it would provide additional support for the claim that the implementation of vibrotactile in-car displays for directional information presentation represents a potentially practical and useful means of preventing the front-to-rear-end accidents that often result from a driver being distracted (see McEvoy et al., 2007a; Strayer and Drews, 2004). Experiment 5.3 was also conducted in order to examine whether there could be any 'intuitive' mapping of certain driving actions to a new class of vibrotactile warning signals.

Driving simulator experiment

Eleven participants (mean age of 30 years) drove in a high-fidelity driving simulator at the TRL (Transport Research Laboratory) in Wokingham, England, in which a car following scenario was modelled. The participants had to try and maintain a safe headway distance (2 s time headway) to the lead vehicle using the aid of an in-car navigation tri-box headway distance indicator. This tri-box indicator acted as an analogue to the visually-demanding in-car technology that is distracting drivers' attention (e.g., Ashley, 2001).

As the participants drove through a simulated rural area, they had to respond as quickly as possible to randomly-presented sudden deceleration of the lead car, which in this experiment happened to have its brake lights disabled. All of the participants took part in different blocks of trials either with or without the aid of vibrotactile warning signals that indicated the likely direction of any potential collision events. The vibrotactile warning signals were presented to either their front or back, using the same vibrotactile belt used for the previous two experiments. The 500 ms vibration used in this experiment was spatially-predictive, with the cues correctly indicating the direction of the potential collision event on 80 per cent (as in Experiment 5.1) of the trials (see Ho et al., 2006a, for a detailed description of the methodology).

The results of our driving simulator study revealed that drivers were able to brake significantly more rapidly (348 ms faster) and stopped with a larger safety margins (3.5 m greater stopping distance) when the vibrotactile warning signal was presented than when it was not (see Figure 5.4). Our results revealed a 24.7 per cent reduction in response latency and a 33.7 per cent increase in safety margin to a potential collision when a valid vibrotactile cue that indicated the correct direction of the critical event was presented than when no cue was given. The results of Experiment 5.3 also revealed a 13.6 per cent reduction in the latency of our drivers' responses when the vibrotactile cue was invalid (i.e., the cue did not predict the correct direction of the upcoming

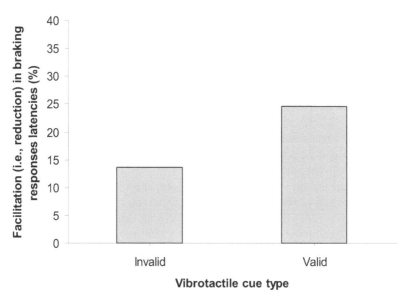

Figure 5.4   **Graph highlighting the facilitation of braking latencies reported in the simulator-based driving experiment associated with the use of valid (i.e., indicating the likely direction of the critical driving event) and invalid vibrotactile warning signals when compared to when no warning signal was presented, as is common in many cars today (see Ho et al., 2006a, for details)**

critical event). Taken together, the results of Experiment 5.3 highlight the important safety implications that the introduction of tactile warning signals may have for real-world driving performance (see Brookhuis et al., 1994; Suetomi and Kido, 1997).

The results of our simulator study further reinforce the idea of the potential benefits offered by directional vibrotactile cues in eliciting a braking response from a driver in an applied setting. In particular, our results highlight the efficacy of using this seemingly 'intuitive' vibrotactile display to present time-critical information to drivers. It is important to note that the term 'intuitive' has been used by many researchers when referring to the presentation of information via the tactile channel (e.g., Eves and Novak, 1998; Rupert, 2000b; van Erp, 2005; van Erp and van Veen, 2004). The concept of 'intuitive' encompasses a straight-forward transfer of tactile information to/from the brain that presumably bypasses the traditional routes in information-processing to elicit the most rapid response from an operator.

The results of Experiment 5.3 clearly demonstrate that the presentation of valid frontal vibrotactile cues offers an effective means of eliciting a braking response from a participant/driver who has had no prior training with the vibrotactile system. The study by Jones et al. published in 2008, (see Jones, Gray, Spence, and Tan, 2008) also showed that localized spatially-informative vibrations presented to the back of their participants can be used to direct a person's visual attention to one of the forward-facing quadrants of the visual field. Given the robust benefit attributable to the provision of vibrotactile cues under normal conditions, it seems plausible that even bigger benefits of vibrotactile cuing will be evidenced in situations that are more visually attention-demanding or degraded, such as in low visibility or heavy traffic, when drivers are under high perceptual load (see Janssen, Michon, and Harvey, 1976; Lavie, 2005). However the one potential caveat to our findings was mentioned earlier, namely that drivers in this driving simulator study are presented with warning signals at a rate far more frequent than is likely to be the case in real life, should not be forgotten (see Chapter 8 for a fuller discussion of this issue).

With a growing population of elderly drivers, the communication of redundant information via touch is also promising as studies have demonstrated greater performance gains in the elderly population than the younger population when presented with multisensory as opposed to unimodal stimuli (see Laurienti, Burdette, Maldjian, and Wallace, 2006; Spence and Ho, 2008). Furthermore, a large body of empirical research has now shown that older drivers typically find it harder to dual task effectively (i.e., safely) when behind the wheel than do younger drivers (e.g., see Alm and Nilsson, 1995; Hancock et al., 2003; McKnight and McKnight, 1993; Reed and Green, 1999; Shinar et al., 2005; Violanti, 1997; though see Briem and Hedman, 1995; Jenness et al., 2002; McEvoy et al. 2005, for null results).

The results of Experiment 5.3 demonstrate the potential benefits of using 'intuitive' vibrotactile in-car displays as a means to present time-critical information to drivers other than by the visual and auditory driving aids that are currently widely used. The findings demonstrate the applicability of the insights derived from the theoretical account of crossmodal links in spatial attention between various different sensory modalities (as outlined in Chapter 1) in a practical application domain. The results also provide further support for the significant spatial cuing effects reported in the earlier laboratory-based studies covered in this chapter. Taken together, these findings add to a growing body of empirical evidence highlighting the potential benefits associated with the use of 'intuitive' vibrotactile in-car displays in order to alert drivers to potential front-to-rear-end collisions and to provide time-critical directional information. Future studies should investigate the speed with which tactile information is translated into signals for decision-making and subsequent actions in order to have a better understanding of the mechanisms governing the seemingly spontaneous and automatic translation of a vibrotactile stimulus on the stomach to a braking response (see Proctor et al., 2005; Spence and Ho, submitted). The development of vibrotactile warning signals that have been specifically designed to play to the particular strengths of the human information processing system would therefore seem to offer some exciting possibilities for future car warning signal design.

Chapter 6

The Multisensory Perceptual versus Decisional Facilitation of Driving

Introduction

The potential use of non-visual warning signals to present spatial information to car drivers has been successfully demonstrated by the various experiments reported in Chapters 4 and 5 (for the case of unimodal auditory and unimodal tactile warning signals, respectively). The empirical research reported in the two preceding chapters therefore suggests that spatial warning signals can be used to improve a person's ability to detect, and subsequently to respond to, events occurring in the direction from which the cue, or warning signal, is presented. Among the two types of spatial warning signals investigated (namely, auditory and vibrotactile), spatial auditory cues were found to be particularly effective in directing a driver's visual spatial attention to potentially dangerous events on the road. Such warning signals should hopefully be particularly useful in reorienting the attention of a driver who has been distracted by a secondary task, such as talking on the mobile phone or to a passenger. However, given that the spatial cuing effects were measured at the behavioural level, it is unclear what the factors are that govern the difference in the relative effectiveness of the two types of cue. The experiments reported in this chapter were therefore designed to attempt to understand the level(s) of information processing at which crossmodal cuing effects attributable to the spatial warning signal occur.

Relative warning signal effectiveness across sensory modalities

One important factor to consider when comparing the effectiveness of various different classes of warning signal is how long it will take people to respond to them, given the well-documented differences in transduction latencies for stimuli presented in different sensory modalities (e.g., Harris, Harrar, Jaekl, and Kopinska, forthcoming; Spence and Squire, 2003). The available research suggests that people can respond more rapidly to tactile stimuli presented to their hands than to visual stimuli (e.g., see Spence et al., 2001). However, given that reaction times are likely to be slower for vibrotactile stimuli presented to the torso (than to the fingertip, as studied in the majority of previous studies), this is an important question that needs to be addressed (see Bergenheim, Johansson, Granlund, and Pedersen, 1996; Harrar and Harris, 2005).

A preliminary comparison of the mean reaction times from the experiments described in Chapters 4 and 5 would appear to suggest that when people are multitasking, they can respond more rapidly to visual events following the presentation of an auditory warning signal than following the presentation of a vibrotactile warning signal (see Figure 6.1). Experiment 6.1 was therefore conducted in order to compare the speeded discrimination of the warning signals used in these earlier experiments and so examine the relative speed with which people can respond to auditory, tactile and visual warning signals (see Ho, Spence, and Tan, 2005a). The hope here was that the results of Experiment 6.1 would enable comparison of the relative effectiveness of the various warning signals across different sensory modalities. It seemed possible that the differences in response latency are primarily governed by the speed with which people can respond to warning signals presented in a given sensory modality. Alternatively, however, the differences could also be determined by the relative

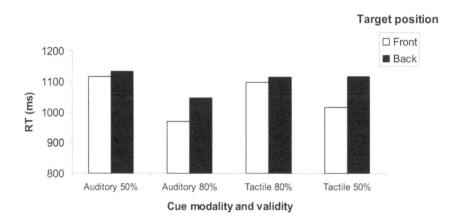

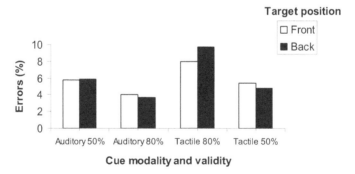

Figure 6.1   **Summary graphs highlighting (top) mean reaction times (RTs; ms) and (bottom) percentages of errors as a function of the visual target position (front vs. back) reported in Experiments 4.2 (Auditory 50 per cent), 4.3 (Auditory 80 per cent), 5.1 (Tactile 80 per cent) and 5.2 (Tactile 50 per cent), investigating the effectiveness of various warning signals presented in different sensory modalities**

efficacy with which people can 'translate' from the detection of the warning signal itself to looking at the appropriate visual event(s).

The participants in Experiment 6.1 were instructed to respond as soon as possible when they detected a warning signal presented from one of two possible positions, presented either auditorily, tactually, or visually (in separate blocks of trials) from one of two possible positions (see Figure 6.2). In particular, we wanted to assess whether the spatial location from which the target warning signals were presented had any effect on the participants' ability to detect them, and whether the detection of, and subsequent responses to, warning signals presented in the three different sensory modalities differed.

The results of Experiment 6.1 demonstrated that our participants were able to respond to vibrotactile warning signals more rapidly than to either visual or auditory warning signals (see Figure 6.3).[1] Given the increasing visual overload reported by interface operators in many interface settings (e.g., Sorkin, 1987; Spence and Driver, 1997a), the use of non-visual warnings may seem more effective than visual warning signals in alerting a person engaged in a primarily visual task such as driving (e.g., Morris and Montano, 1996; Sorkin, 1987; though see also Sivak, 1996). It should though be noted that some researchers have actually suggested that visual overload may arise not only from human operators having to deal with too much information, but also from a lack of knowledge or awareness regarding which of the available sources of visual information is actually most relevant at any given point in time (e.g., Perrott et al., 1996).

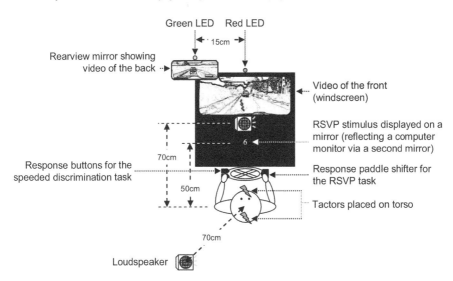

Figure 6.2   Schematic diagram of the experimental set-up used in Experiment 6.1

---

1   Note that the differences between the various sensory modalities reported here were governed not only by any differences in transduction latencies but also by other factors such as the distance from the source stimuli (e.g., the vibrotactile stimuli were presented on the body, while the auditory and visual information originated from a source 70 cm from the body), and the specific parameters (e.g., intensity, stimulus duration) chosen.

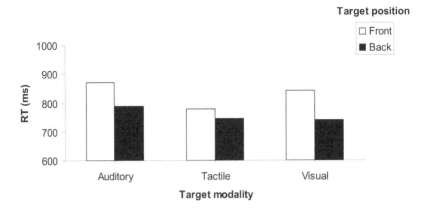

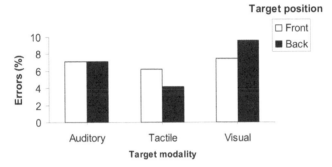

Figure 6.3   **Summary graph highlighting (top) mean reaction times (RTs; ms) and (bottom) percentages of errors in the speeded target discrimination task in Experiment 6.1 as a function of target modality and position**

The participants in Experiment 6.1 reacted significantly more rapidly, and somewhat more accurately, to vibrotactile target stimuli delivered to their torso than to auditory stimuli presented a short distance from their ears. In fact, the use of vibrotactile cues in an automobile setting may be more advantageous than the use of auditory cues not only because people can react to them more rapidly, but also because they are private to the driver, so other car passengers should not be disturbed by their occurrence (see Chapter 5; Triggs et al., 1974; van Erp and van Veen, 2001, 2004, for a discussion of the potential use of vibrotactile displays in automobiles and aeronautical applications). Vibrotactile cues also have the advantage that they should not be affected by the level of background auditory noise in the driving environment to the same extent as are auditory cues (Gescheider and Niblette, 1967; although, of course, the ability to detect vibrotactile stimuli might be masked somewhat by any vehicular vibrations; cf. Kemeny and Panerai, 2003).

One reason for the slow discrimination performance reported in response to the auditory targets in Experiment 6.1 may be related to the fact that people find it more difficult to localize sounds within confined spaces such as in a car (e.g., Catchpole et al., 2004; Fitch et al., 2007; Moore and King, 1999). Psychophysical studies of tactile perception, on the other hand, have shown that participants can accurately localize vibrotactile stimuli presented at up to twelve different sites on their abdomen, with localization accuracy being primarily determined by how close the vibrotactile stimuli are to the spine or navel, and how close any two given sites are to each other (Cholewiak et al., 2004). Fitch et al. (2007) have also recently shown that drivers can rapidly discriminate which of eight directions were signalled by the vibration of the driver's seat. Such vibration, in fact, proved to be far more effective than auditory directional signals. Future research should consider whether there are other body sites (perhaps closer to the driver's brain; cf. Bergenheim et al., 1996; Gilliland and Schlegel, 1994; Harrar and Harris, 2005) where vibrotactile reaction times would be even faster than those observed in Experiment 6.1, given that the waist may not be the optimal site for presenting vibrotactile stimulation (e.g., see Ho and Spence, 2007; Weinstein, 1968).

The results of Experiment 6.1 allow for a basic comparison to be made between the results of the previous experiments on the auditory and tactile spatial cuing of driver attention. The results show that the efficiency with which participants can respond to dangerous road events signalled by auditory, visual, or vibrotactile warning signals depends not just on the speed with which participants can discriminate the position from which a warning signal has been presented. For instance, in contrast to the faster discrimination latencies seen for vibrotactile rather than auditory warning signals in Experiment 6.1, participants in the spatially-predictive auditory experiment (Experiment 4.3) were, on average, numerically somewhat faster ($M = 1008$ ms) overall than participants in the spatially-predictive vibrotactile experiment (Experiment 6.1; $M = 1104$ ms). The contrasting pattern of results reported in these two experiments would therefore appear to suggest that the cue and target modalities may interact to determine the resultant effectiveness of any particular warning signal in a real-world setting. Furthermore, there are also costs associated with the shifting of attention from one sensory modality (that of the warning signal) to another (i.e., that of the critical driving event seen visually) that also need to be taken into account (Rodway, 2005; Spence and Driver, 1997a; Spence et al., 2001; Turatto, Benso, Galfano, Gamberini, and Umiltà, 2002; Turatto, Galfano, Bridgeman, and Umiltà, 2004) when considering the modality in which it will be most appropriate to introduce a given warning signal.

Perceptual vs. decisional facilitation

The rapidly-growing body of empirical research on the nature of crossmodal links in spatial attention demonstrates that responses to target events presented in one sensory modality can be facilitated by the prior presentation of a relevant cue in another sensory modality from approximately the same spatial location (see earlier chapters; Spence and Driver, 2004, for a recent review). Such crossmodal spatial facilitation effects have been studied in a variety of different settings, from fundamental laboratory-based research (e.g., Driver and Spence, 2004; Spence et al., 2004) to more complicated environments in applied research (e.g., Fitch et al., 2007; van Erp and van Veen, 2001, 2004). However, it is unclear whether the facilitatory effects reported in many of these previous cuing studies should be attributed to response priming, to attentional facilitation or to some unknown combination of these two effects.

According to the response priming account, the facilitation of a participant's performance seen following the presentation of a warning signal (or cue) occurs because the response appropriate to the target stimulus is primed by the cue (i.e., people's performance is facilitated without there necessarily being any change in the perceptual representation of the target itself). According to the attentional account, however, the presentation of the cue (or warning signal) leads to a crossmodal shift of a person's spatial attention that facilitates the perception of targets subsequently presented near to the cue. Such attentional enhancement of target processing is what gives rise to improved behavioural performance. The critical research question to be addressed here then is one of understanding how 'early' in human information processing the crossmodal cuing effects attributable to the presentation of a warning signal occur.

From an applied perspective, the facilitation of behavioural responses to potential emergency situations is clearly beneficial no matter how that benefit is elicited (i.e., no matter whether it is elicited by response priming, attentional facilitation, or by some unknown combination of these two effects). From a theoretical point of view, however, many researchers have undertaken extensive laboratory-based research specifically in order to try and isolate response priming (or compatibility) effects while others have focused on trying to isolate attentional facilitation effects instead (see Proctor et al., 2005; Spence and McDonald, 2004). It is only by gaining a better understanding of the relative contributions of the two effects (response priming and attentional facilitation) to performance under various different more realistic conditions that it might, in the future, be possible to design the most effective warning signals – signals that will presumably facilitate performance at both levels simultaneously.

In an attempt to eliminate the possible contribution of response priming in laboratory-based studies of crossmodal cuing, Spence and Driver (1994) proposed the use of an orthogonal cuing methodology. In a typical orthogonal cuing study, participants are instructed to respond to a dimension, or property, of the target (e.g., its colour or identity) that is orthogonal to the dimension along which cuing occurs. So, for instance, if participants have to make red versus blue target discrimination responses for targets appearing on either the left or right, then the spatial cuing dimension (left vs. right) is orthogonal to the dimension on which responses are made (red vs. blue). The idea is that the presentation of the cue should therefore not prime a particular response, and so any residual facilitation of performance can unambiguously be attributed to attentional cuing. Thus, the dimension to which the participants need to respond has to be orthogonal to the dimension of interest of the experimenters.

Taken together, the series of simulated driving experiments reported in Chapters 4 and 5 have demonstrated that the use of various different types of spatial auditory and vibrotactile warning signals (presented from either the front or rear) can facilitate performance when responding to targets presented from the front or rear. In these experiments, the participants were required to brake or to accelerate in response to target visual driving events seen through the front windscreen or in the rearview mirror respectively (i.e., as required in a real driving situation). However, it is unclear whether the facilitation of responding in these previous studies reflects response priming, spatial attentional facilitation or some unknown combination of the two effects. The experiments reported in this chapter were therefore designed to study whether a purely attentional component to this spatial warning signal effect could be demonstrated by requiring participants to perform an *orthogonal* discrimination task (see Ho, Tan, and Spence, 2006b). Specifically, our participants had to make speeded discrimination responses regarding the *colour* of a numberplate (i.e., responses that were orthogonal to the spatial dimension of interest – the front vs. back cuing of driver attention), with the experiment otherwise being identical to those reported in the previous experimental chapters. The results of the experiments reported in this chapter should therefore highlight those factors governing the facilitation effects reported in the earlier chapters.

It was hypothesized that if the same pattern of results could be replicated using an orthogonal discrimination task as in the previous experiments, then this would suggest that the facilitatory crossmodal cuing effects observed previously must reflect, at least in part, a spatial attentional effect. Alternatively, however, if the cues (warning signals) that were presented in the previous experiments simply primed the appropriate responses (i.e., braking vs. accelerating), then no such facilitation would be expected in the orthogonal experiments (given the orthogonal nature of the responses required). Experiment 6.2 was designed to replicate Experiment 5.1 where spatially-predictive vibrotactile cues (80 per cent valid with regard to the likely location of the upcoming visual targets) had been shown to facilitate subsequent responses to target visual driving events in the cued direction. Experiment 6.3 replicated Experiment 4.3, where spatial auditory car horn sounds were used as the spatially-predictive warning signals.

Somewhat surprisingly, the results of Experiment 6.2 failed to reveal any spatial cuing effect following the presentation of the vibrotactile cues (see Figure 6.4A). In fact, there was no significant difference in the latency of participants' responses as a function of whether the vibrotactile cues came from the same or opposite direction as the critical visual driving targets. The slightly faster responses that were observed following a cue from behind as compared to following a cue delivered from the front may reflect the possibility that vibrotactile stimuli are more salient (and/or alerting) when presented to the back than to the front of a participant's waist.

The null vibrotactile cuing effects reported in Experiment 6.2 contrast with the robust cuing effects reported in the earlier study of vibrotactile cuing of driver attention. Taken together, these results therefore suggest that while the presentation of a vibrotactile cue from the appropriate spatial direction on a person's body surface may elicit an automatic (or 'intuitive'; van Erp, 2005) response bias (cf. Prinzmetal, McCool, and Park, 2005; Spence et al., 2004), it does not necessarily lead to a shift of spatial attention (that would have been expected to lead to a perceptual facilitation of colour discrimination responses).[2]

---

2   It could be argued that spatial attention might not have influenced the detection of colour changes and the detection of angular size changes (as in the detection of a rapidly-approaching car reported in the experiments outlined in Chapters 4 and 5) in the same way, thus rendering a strict comparison between the present study and the previous studies somewhat problematic. However, it should be noted that whether the spatial element is inherent in the task itself (e.g., as in the angular size change detection task) or not (e.g., as in the colour change detection task) is a constant factor within each of the experiments, and that any effects attributable to this factor should be cancelled out when the mean cuing effects within each experiment are calculated. Note that the mean cuing effects reflect the effectiveness of the spatial cues to

The Multisensory Driver

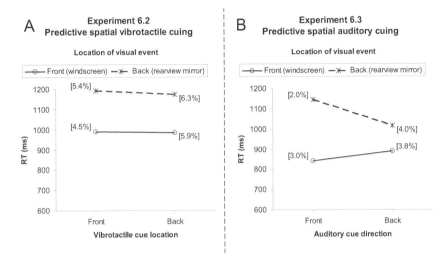

Figure 6.4   Summary interactions in the mean reaction time (RT; ms) data between the direction of the cue and the location of the critical visual event in the orthogonal task in Experiments 6.2 (A) and 6.3 (B). Percentages of errors are indicated by numerical values in parentheses

The results of Experiment 6.2 show that the presentation of a vibrotactile cue to the front or back of the participant's torso did not lead to a shift of their spatial attention in the direction cued by the vibration as indexed by a task involving the discrimination of the colour of a numberplate presented far from the participant's body. It is possible, however, that the vibrotactile cues might have resulted in our participants' attention being oriented toward the source of stimulation on their torso (i.e., toward peripersonal space; see Kitagawa, Zampini, and Spence, 2005; Kitagawa and Spence, 2006; Tassinari and Campara, 1996), rather than to the extrapersonal visual space outside the car (as seen through the windscreen or rearview mirror). Previous research has shown that crossmodal spatial cuing effects are maximal when the cue and target are presented from the same spatial *location*, and decline as the cue-target separation increases (Spence et al., 2004). Note that this contrasts with the case of response priming, where as long as the cue primes the appropriate response, it does not really seem to matter so much what the exact spatial relationship between the cue and target is; see Chapter 7 on this point). Indeed, recent neuropsychological research with people suffering from unilateral spatial extinction (these people fail to perceive or respond to stimuli in the space contralateral/opposite to their brain damage) has shown far more pronounced interactions between pairs of stimuli

---

orient the attention of the participants to the spatial location of interest. Other researchers have also reported the same pattern of attentional orienting effects across experiments no matter whether the task was to detect for a target presented alone or amongst distractors, or else to discriminate the colour of a target following the presentation of the same visual cues (e.g., see Hommel, Pratt, Colzato, and Godijn, 2001).

when they are both presented within the same region of space (i.e., both within peripersonal space or both within just extrapersonal space) than when they occur in different functional regions of space (i.e., one in peripersonal space and the other in extrapersonal space; e.g., Làdavas, 2002). Such results have been taken to show that the brain represents stimuli originating in these two regions of space in a somewhat distinct manner (cf. Previc, 1998, 2000; see also Holmes and Spence, 2006).

Given that vibrotactile cues can only be presented on the body surface itself (i.e., in near-peripersonal space), Experiment 6.3 was designed to investigate whether the presentation of spatially-predictive auditory warning signals from the same functional region of space as the target visual driving events (i.e., when both events were presented in extrapersonal space) would facilitate performance on the numberplate colour discrimination task for targets presented from the cued location. (Note that both the rear auditory and visual stimuli originated from the rear, despite the fact that the targets were visually inspected from *in front* via the rearview mirror.)

The participants in Experiment 6.3 were found to respond significantly more rapidly to target visual driving events (the sudden change in the colour of a car's numberplate) occurring in the same direction as the auditory warning signal than when they were presented from opposite directions (see Figure 6.4B). These results therefore replicate the crossmodal spatial facilitation effects observed in the previous auditory cuing study when participants either had to brake or to accelerate in response to their sudden approach toward the car in front or the rapid approach of the car behind (see the results of Experiments 4.1–4.5). These results also allowed us to rule out response priming as the sole explanation for the crossmodal effects in the auditory cuing experiments reported previously. Taken together, these findings support the idea that informative auditory warning signals can effectively orient a driver's spatial attention to a particular region of space (in this case, to front or rear extrapersonal space), leading to a facilitation of responses to visual events occurring subsequently in that direction (e.g., Spence et al., 2004).

Differential effects of auditory and vibrotactile cuing

The facilitatory effects of crossmodal spatial attentional cuing were observed following the presentation of spatially-predictive auditory warning signals (Experiment 6.3), but not following the presentation of spatially-predictive vibrotactile warning signals (Experiment 6.2). These findings contrast with the significant crossmodal spatial facilitation effects evidenced following the presentation of both spatially-predictive auditory and vibrotactile warning signals that were reported earlier (see Figure 6.5). There are at least two possible explanations for the differential effects of auditory and vibrotactile cuing on performance in the experiments reported in this chapter. First, it may be that audition and touch affect visual spatial attention in somewhat different ways. For although it has been argued previously that spatial attention may function in the same manner across the various 'spatial' sensory modalities (see Spence and Driver, 2004), the translation (or switching of attention) from touch to vision has been shown to be less efficient (i.e., more time-consuming) than from audition to vision (e.g., see Spence et al., 2001). The change in the colour of the numberplate (which was superimposed on

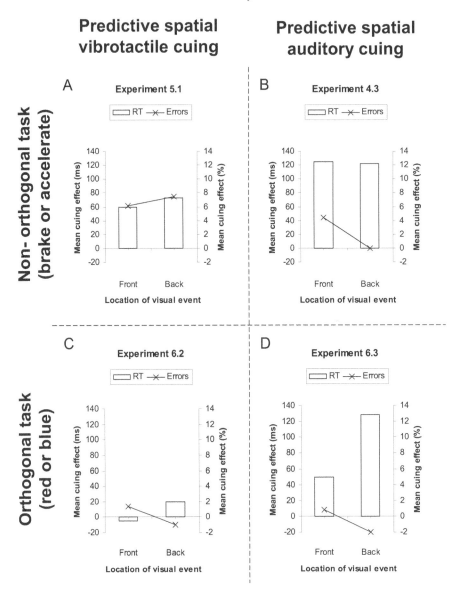

Figure 6.5   Summary graphs highlighting the mean cuing effects following
              vibrotactile (A) and auditory (B) cuing in the non-orthogonal
              driving task in Experiments 5.1 and 4.3, respectively, and
              vibrotactile (C) and auditory (D) cuing in the orthogonal task in
              Experiments 6.2 and 6.3, respectively

a rapidly approaching, i.e., looming, car) in the orthogonal discrimination experiments reported in this chapter may simply have been more attention-capturing (cf. Franconeri, Simons, and Junge, 2004; Turatto and Galfano, 2001) than the rapid approach toward the car in front or the rapid approach of the car from behind in the previous experiments (though see Terry, Charlton, and Perrone, forthcoming). Consequently, the change of the colour of the numberplate may have more automatically (and/or rapidly) captured the attention of our participants in Experiments 6.2 and 6.3. As tactile information is transduced at the receptor surface (i.e., skin) more slowly than auditory stimuli at the basilar membrane of the ear (see Spence and Squire, 2003; Zampini, Brown, Shore, Maravita, Röder, and Spence, 2005), it might be suggested that the participants in Experiment 6.2 did not have sufficient time to localize the vibrotactile stimuli and shift their spatial attention to the visual driving task from the RSVP task prior to their attention being captured automatically by the onset of the visual target (Experiment 6.3), and so not following the presentation of spatially-predictive vibrotactile warning signals (Experiment 6.2). These findings contrast with the significant crossmodal spatial facilitation effects evidenced following the presentation of both spatially-predictive auditory and vibrotactile warning signals in the previous experiments (see Figure 6.5).

However, the results of Experiment 6.1, where the relative speed with which people could discriminate the direction (front vs. back) from which warning signals were presented in the different sensory modalities was investigated directly, reveals this explanation to be incorrect. In that experiment, it was found that participants could discriminate the direction of vibrotactile warning signals significantly more rapidly than the direction of auditory warning signals, presumably because of the difficulty people have in localizing auditory stimuli within confined spaces (e.g., Catchpole et al., 2004; Fitch et al., 2007; Moore and King, 1999).

Alternatively, however, it could also be argued that the colour discrimination task itself may simply have been less sensitive to the distribution of our participants' spatial attention than the looming (i.e., braking/accelerating) task used in the previous experiments, with the latter task in some sense involving the integration of spatial information over time (cf. Spence et al., 2004). There are two reasons why such an account can be ruled out: First, several researchers have previously shown reliable spatial attentional orienting effects using colour discrimination tasks (e.g., see Hommel et al., 2001). Second, the reliable spatial cuing effects observed with auditory cuing in Experiment 6.3 argue against the colour discrimination task simply being insensitive to the distribution of our participants' spatial attention. Hence, it is very unlikely that the use of the colour discrimination task represents a major factor contributing to the null vibrotactile cuing effect observed in Experiment 6.2.

A number of researchers have suggested that the use of vibrotactile cues may provide an 'intuitive' means of presenting directional information in interface displays, such as in situation-awareness and navigation systems for pilots and drivers (e.g., Rupert, 2000a; van Erp, 2005; van Erp and van Veen, 2004). For instance, van Erp and van Veen showed that drivers could respond more rapidly following navigational messages presented in a bimodal (tactile and visual) display than when the messages were presented unimodally (i.e., via touch or vision only). Participants also reported significantly lower subjective mental workload ratings with the touch-only display, as compared to when using the visual-only or bimodal displays (see also Fitch et al., 2007; Lee and Spence, forthcoming).

More recently, Van Erp (2005) reported that people seem to have no trouble in localizing vibrotactile stimuli presented to their torsos, and can indicate the associated direction in the external environment accurately (albeit with a consistent mislocalization bias toward the navel or spine). However, the results of Experiment 6.2 suggest that vibrotactile cues are in some sense limited. Indeed, when assessing the effectiveness of vibrotactile cues in conveying information, it may be useful to break the role of such stimuli into distinct categories, namely, their ability to direct spatial attention to the location of the critical target event, their ability to prime the response appropriate to that target event, and also their ability to act as a means of information transfer to an interface operator.

The most likely explanation for the difference in the effectiveness of auditory and vibrotactile cuing in Experiments 6.2 and 6.3 comes from recent evidence suggesting that the brain represents stimuli occurring in peripersonal space somewhat differently from those occurring in extrapersonal space (e.g., Kitagawa et al., 2005; Previc, 1998, 2000; Rizzolatti, Fadiga, Fogassi, and Gallese, 1997; Spence and Driver, 2004; Tajadura-Jiménez et al., submitted). It may be that the remapping of a front or back directional cue on the torso (i.e., in peripersonal space) to a target visual event taking place in a location farther away in extrapersonal space is less efficient than when the cue and target both occur within the same functional region of space (i.e., both within peripersonal space or both within extrapersonal space). In this regard, it would be interesting in future research to examine how people subjectively represent extrapersonal and peripersonal space while they are driving (e.g., whether the peripersonal vs. extrapersonal distinction equates to events happening within the car vs. outside the car, respectively, cf. Kitagawa and Spence, 2005; and whether the region of protective personal space expands under conditions where an individual feels threatened, e.g., Dosey and Meisels, 1969; Felipe and Sommer, 1966). Understanding the representation of near-peripersonal space may therefore assist us in interpreting the present findings (see Kitagawa et al., 2005; Làdavas and Farnè, 2004).[3]

If, as recent cognitive neuroscience data suggests, the brain really does deal with peripersonal and extrapersonal space in qualitatively different ways (Làdavas, 2002; Previc, 1998; Spence and Driver, 2004; Weiss et al., 2000), then it might be predicted that the crossmodal spatial facilitation effects observed in Experiment 6.3 would be eliminated if the spatial auditory cues were actually presented from sources that were much closer to the participants' body (just as was the case for the vibrotactile cues used in Experiment 6.2). If, however, crossmodal spatial facilitation effects were still to be demonstrated in this situation, then it would suggest that there must be another cause for the differential effects of auditory and vibrotactile cues on visual spatial attention in the driving situation. Experiment 6.4 was therefore designed to investigate whether the presentation of the same spatial auditory cues used in Experiment 6.3 from locations closer to the participants' body would, in fact, eliminate the crossmodal spatial facilitation effects observed previously (see Figure 6.6).

---

3   It would also be interesting to examine what role, if any, the fact that a driver is moving rapidly in a predictable direction has on the shape and border of peripersonal versus extrapersonal space (cf. Gibson and Crooks, 1938).

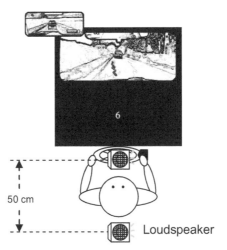

Figure 6.6   Schematic diagram of the experimental set-up used in Experiments 6.4 and 6.5

Consistent with the hypothesis derived from recent cognitive neuroscience research suggesting that the brain represents the various functional regions of space surrounding the body somewhat differently, no spatial cuing effect was observed following the presentation of spatial auditory cues when the loudspeakers presenting the auditory cues were moved from the extrapersonal space far away from the body (coinciding with the location of the monitor in which the visual stimuli were seen) to locations closer to the participant's own body (in peripersonal space, as in the case of the vibrotactile stimuli, even though the sources of the auditory cues were not actually touching the participant's body as such; see Figure 6.7A). Taken together, these findings suggest that the perceptual/attentional enhancement of the representation of a target event may require the presentation of a cue from the same functional region of space as the target, regardless of the sensory modalities in which the cue and target are presented.

One potential confound, however, with the set-up utilized in Experiment 6.4 was that the placement of the two auditory sources was susceptible to potential front-back confusions (Geissler, 1915; Spence et al., 2000b; Stevens and Newman, 1936), thus limiting the spatial property of the auditory cues (meaning that the condition would resemble more a nonspatially-predictive auditory cuing case, as in Experiment 4.2. Note though that a significant spatial cuing effect was found in Experiment 4.2 with the presentation of 50 per cent valid auditory car horn sounds from loudspeakers located far away from the participant's body). In order to address this issue, a replication of Experiment 6.4 was conducted with the use of white noise bursts instead of auditory car horn sounds. The results of previous studies have suggested that participants find it easier to localize white noise bursts than pure tones presented from proximal sources (Deatherage, 1972; Spence and Driver, 1994; Stevens and Newman, 1936).

Experiment 6.5 was designed to investigate whether the null effect of spatial cuing following the presentation of auditory cues from close to the body was an artefact of any difficulty that our participants may have had in localizing the spatial

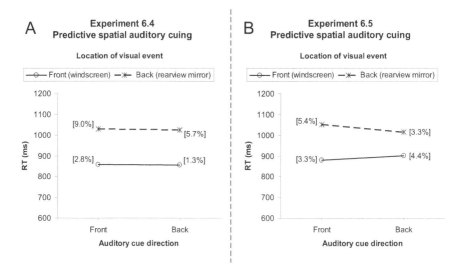

Figure 6.7   Summary interactions in the mean reaction time (RT; ms) data between the direction of the auditory cue and the location of the critical visual event in the orthogonal task presented from close to the driver's head in Experiments 6.4 (car horn sounds; A) and 6.5 (white noise bursts, B). Percentages of errors are indicated by numerical values in parentheses

auditory car horn sounds. In fact, the results of Experiment 6.5 failed to show any significant spatial cuing effect for auditory white noise bursts presented from sources close to the head of the participants, once again providing evidence to support the functional region of space hypothesis (see Figure 6.7B). Between-experiments analyses were conducted in order to compare the spatial cuing effects found in the three orthogonal auditory cuing experiments. A between-experiments analysis of the mean cuing effect in the reaction time data obtained in Experiments 6.3–6.5 revealed a significant difference (see Figure 6.8). Post-hoc comparisons further showed a significant difference between spatially-predictive auditory car horn sounds presented far from the head and those presented from close to a participant's head, and between spatially-predictive auditory car horn sounds presented far from head and spatially-predictive auditory white noise bursts presented close to head, but not between the two spatially-predictive auditory cues both presented close to head.

Overall, the results of the five experiments described in this chapter together suggest that in order for a spatial multisensory warning signal to effectively enhance a person's perceptual representation of the cued spatial location, the cue has to be presented from the same functional region of space as the target event. Otherwise, a spatial cue may only inform a person with regard to the potential target direction and prime the appropriate response, rather than leading to any attentional shift to the cued direction. Both vibrotactile and auditory cues presented on/close to the torso in personal/peripersonal space were simply not found to be capable of enhancing the perceptual saliency of a visual target event presented in extrapersonal space away from

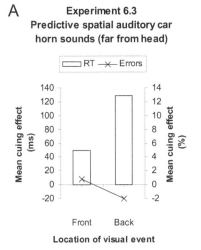

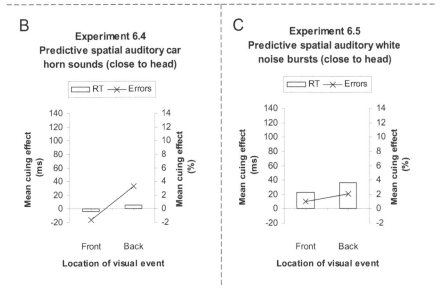

Figure 6.8   **Summary graphs highlighting the mean cuing effects following spatially-predictive auditory car horn sounds presented far from the head (A), spatially-predictive auditory car horn sounds presented from close to the head (B) and spatially-predictive auditory white noise bursts presented from close to the head (C)**

the torso, even though these cues were effective in priming the appropriate behavioural responses in the applied level (see the experiments reported in Chapters 4 and 5).

One current topic of interest in the area of multisensory warning signal design is the relative effectiveness of unimodal versus bimodal and/or multimodal signals. To date, published studies have demonstrated mixed findings with regard to the superiority of multisensory over single channel (or unimodal) information presentation (e.g., Selcon et al., 1995; Spence and Driver, 1999; see also Campbell et al., 1996; Fitch et al., 2007). One important distinction to be made here is the one between those multisensory signals that convey the same information (i.e., that are, in some sense, redundant) versus those that provide information concerning different, perhaps interdependent, aspects of the same event (cf. Fidell, 1982).

With regard to the former situation, a number of studies have demonstrated benefits of redundant sources of information presentation. For instance, in one oft-cited study, Selcon et al. (1995) compared the speed with which participants could detect and respond to the presentation of a number of different combinations of visual and/or auditory spatial and/or verbal missile approach warning signals. They observed significantly faster response latencies in their participants when two to four sources simultaneously gave the same information than when only a single type of cuing was used. Hence, the multisensory warning signals used in Selcon et al.'s study would appear to have resulted in an additive effect on human performance. It should, however, be noted that it is unclear whether the cues facilitated performance simply by priming the correct response and/or whether they also led to a genuine attentional facilitation effect (cf. Miller, 1982; Spence and Driver, 1999).

On the other hand, Spence and Driver (1999) reported bimodal audiovisual warning signals to be no more effective than unimodal auditory signals in capturing a participant's spatial attention in a laboratory-based cuing study where response priming explanations of their results were ruled out (by cuing the participant's attention in a dimension – left vs. right – that was orthogonal to the response dimension along which participants were required to respond – up vs. down).

It is perhaps worth noting one limitation with the majority of previous studies of crossmodal attentional capture, namely that they have typically assessed the ability of a spatial cue (presented either unimodally or bimodally) to capture attention under conditions where the participants' only task involved responding as rapidly and accurately as possible to the frequently-presented target stimuli (i.e., under single task conditions; see Spence et al., 2004, for a review). This situation is in many ways quite different from the typical situation in which multisensory warning signals would need to be maximally effective, that is, when an interface operator is engaged in another highly attention-demanding task (see Chapters 2 and 3), and where the warning signal needs to be sufficiently salient, and attention-capturing, to break through and capture the attention of the driver, who may be intently listening to what someone is saying at the other end of a mobile phone.

There is now a growing body of empirical research comparing the effectiveness of unimodal feedback signals to multisensory feedback signals in a variety of different interface settings (e.g., Akamatsu, MacKenzie, and Hasbroucq, 1995; Cockburn and Brewster, 2005; Lee and Spence, forthcoming; Vitense et al., 2003). For example, Akamatsu et al. have shown that the use of combined tactile, auditory and visual feedback signals in a mouse-pointing task was no more effective than the unimodal tactile signal when presented by itself. Their results presumably reflect the fact that an operator may select the most appropriate sensory channel for feedback depending on the task at hand (see also Burke et al., 1980; Cockburn and Brewster, 2005).

By contrast, Lee and Spence (forthcoming) have recently shown in a simulated driving task, that drivers who have to avoid potential accidents on the roadway ahead while at the same time trying to operate a touch-screen device (e.g., a mobile phone) were able to react more rapidly to the movements of the car in front when given tri-modal feedback (consisting of tactile feedback from the touch-screen, together with visual feedback from the screen, and auditory feedback from a loudspeaker placed just behind the screen), than when given either unimodal visual or bimodal (visuotactile or audiovisual) feedback in response to their button-presses. What is more, the participants in this task also rated their subjective workload as being significantly lower (as measured by the NASA-TLX) in the multimodal feedback condition as compared to the unimodal feedback condition.

Taken together, therefore, it would seem likely that the potential benefit or costs/ interference attributable to the introduction of multisensory warning signals depends to a large extent on the nature of the task being performed. For instance, in a study by Hirst and Graham (1997) on braking responses following the presentation of collision avoidance warnings, a lower incidence of collisions and increased time-to-collision braking scores (calculated by dividing the inter-vehicle distance by the relative speed, i.e., a higher score indicates better performance) were reported for the combined use of an abstract head-up visual display with either an auditory 500 Hz tone or a speech warning ('danger ahead'), instead of the combined use of a pictorial visual cue together with either the tone or speech cue, but no difference in overall effectiveness was reported as a function of whether the visual cue was combined with a non-speech or speech cue.

A possible alternative account of the mixed multisensory facilitation effects reported in previous studies is in terms of the time-window-of-integration model put forward by Colonius and Diederich (2004; see also Miller, 1982; Mordkoff and Egeth, 1993), given that the occurrence of multisensory integration is typically dependent on both the temporal and spatial configuration of the stimuli presented (see also Stein and Meredith, 1993; Stein and Stanford, 2008). Colonius and Diederich's two-stage model predicts that for multisensory signals to be maximally effective, they should be presented close together in time in order for multisensory integration to occur, and that the spatial proximity of the signals will determine the magnitude and sign of this integration effect (e.g., one signal may inhibit the other if the multisensory signals are presented from distinct spatial locations; see also Holmes and Spence, 2005; Stein and Meredith, 1993). This could possibly account for the lack of any multisensory facilitation effect over and above the unimodal spatial cuing reported here. Note also that for multisensory warning signals to be effective, designers should avoid presenting incongruent information even via different sensory channels (e.g., Selcon, Taylor, and Shadrake, 1991).

From an applied perspective, the results reported so far in this book, when taken together, suggest that the use of vibrotactile and auditory warning signals in automobile interface design to inform drivers of the occurrence of time-critical events outside the car is both effective and practical as a means of reducing visual workload while at the same time effectively capturing a driver's attention (and/or priming the appropriate response). The findings that have been reported in this chapter also imply that response compatibility represents an important factor in multisensory interface design, and may facilitate performance whenever the warning signal and the relevant target event are presented from approximately the same direction. However, the additional benefits attributable to attentional facilitation may only occur when the warning signal and target event locations match much more closely (i.e., when they are both presented within the same functional region of space, and/or from the same direction).

Chapter 7

The Multisensory Spatial Cuing of Driver Attention

Introduction

The last decade has seen a number of important developments in the field of interface design. In particular, there has been a very noticeable shift away from the traditional focus on unimodal systems (predominantly visual and auditory) to multimodal (or multisensory) systems that hold the promise of potentially providing users with more integrated or enriched multisensory experiences (e.g., Lee and Spence, forthcoming; Oviatt, 2002; Sarter, 2006) in both the real world (e.g., Mariani, 2001; Tannen et al., 2004) and in virtual environments (e.g., Yamashita, Ishizawa, Shigeno, and Okada, 2005). However, the proliferation of multimodal interfaces has often occurred without a proper consideration of the fundamental limitations, as well as capabilities, of human multisensory information-processing (see Hempel and Altinsoy, 2005; Sarter, 2007; Spence and Driver, 1999). Meanwhile, many researchers have become increasingly interested in the study of the underlying rules that govern how users perceive the various channels of information presented via a multimodal interface. In particular, some researchers have attempted to better understand how to approach the design of the next generation of multisensory interfaces. Consequently, there is an increasing need to refine multimodal design guidelines and principles in order that they meet the needs and actual capabilities of human users (see also Sarter, 2006).

A large body of empirical evidence from laboratory-based behavioural studies and from neurophysiological studies suggests that the simultaneous presentation of stimuli in more than one sensory modality can give rise to multisensory integration effects that may result in an additive or, in certain cases, even a superadditive facilitation of performance (i.e., to a response that is significantly bigger than that predicted on the basis of summing the responses made to the individual signals) relative to when the stimuli are presented unimodally (see Calvert et al., 2004; Laurienti et al., 2006; Stein and Meredith, 1993; Stein and Stanford, 2008; though see also Holmes and Spence, 2005). By contrast, however, should the individual sensory signals not match in some way, then sub-additive responding (i.e., a multisensory response that is worse than the best of the individual responses to the stimuli when presented in isolation) may well occur. At the cellular level, sub-additive multisensory responses are typically seen when the stimuli presented to different modalities are not aligned spatiotemporally (i.e., when they are presented from different spatial locations and/ or at slightly different times). Clearly interface designers and engineers should be striving to develop those multisensory cue combinations that maximize the likelihood of superadditive responding, and avoid those cue combinations that, for whatever reason, may give rise to sub-additive responding.

One of the characteristics that define any potential facilitation resulting from multisensory interaction relates to the spatial positioning of the multisensory stimuli (see Spence and Driver, 2004). Typically, researchers have assessed the spatial cuing effects of particular combinations of cue and target stimuli by measuring the change, if any, in the perceptual salience or behavioural responses to the target stimuli as a function of whether the cue was presented on the same (i.e., cued) or opposite (i.e., uncued) side or location to the target. In particular, previous research has shown that crossmodal spatial cuing effects are maximal when the cue and target stimuli are presented from exactly the same location, with cuing effects declining as a function of increasing cue-target separation (see Spence and Driver, 2004, for a review). Note though that the spatial modulation of cuing effects seems to tap into the exogenous spatial orienting system that seems to be largely independent of those mechanisms of multisensory integration (see Spence et al., 2004, for a discussion of the subtle distinctions between multisensory integration and spatial attention; see also Macaluso, Frith, and Driver, 2001; McDonald, Teder-Sälejärvi, and Ward, 2001; Santangelo and Spence, forthcoming). The importance of the spatial layout of multisensory stimuli has also been demonstrated in applied research, such as, for example, for collision avoidance in driving as described in the earlier chapter.

The findings that have emerged from many applied studies support the potential facilitatory effects offered by multisensory warning signals as compared to unisensory warning signals, particularly in terms of their ability to alert an interface operator to time-critical events that demand immediate attention. For example, Selcon et al. (1995) found that the presentation of redundant (in 2, 3 or 4 dimensions) missile approach warnings spatially via both the auditory and visual channels gave rise to significantly faster performance in the identification of the source direction of the warnings than when only one warning source was presented. Converging evidence has also been reported by van Erp and van Veen (2004) when information about navigation direction (left vs. right) and distance to the upcoming point of directional change (250, 150 or 50 m) was presented via bimodal visuotactile displays rather than when presented solely via a tactile or visual displays in a driving simulator study (see also Doll and Hanna, 1989; though see Fitch et al., 2007; Lee, McGehee, Brown, and Marshall, 2006).

It should be noted that the majority of previous studies that have demonstrated a facilitatory effect of multisensory warning signals have tended to present the component signals in a spatially-congruent manner. This means that the various unisensory components of the multisensory cue were always both presented on the same side as, or on the opposite side to, the target stimuli. One important question arising from these findings is therefore whether or not spatial coincidence is a prerequisite for multisensory facilitation to occur in an applied (as well as in a basic) research setting. For instance, it is interesting to examine whether the presentation of a non-lateralized tactile warning signal (e.g., such as vibrating a driver's seat) can enhance the salience of a spatial auditory warning signal (e.g., a directional beep concerning the source of a potential danger) and thus result in a facilitatory (or even potentially superadditive) effect on performance (e.g., in terms of the orienting of the driver's attention toward the source of danger) that is greater than when the spatial auditory warning signal was presented alone, even when the tactile cue does not, by itself, provide any relevant spatial information concerning the target event (see also Fitch et al., 2007).

In terms of the exogenous orienting of a person's spatial attention, laboratory-based behavioral studies have generally reported that no particular advantage is associated with the presentation of bimodal spatial cues relative to the presentation of unimodal spatial cues, at least when response biases, criterion-shifts and other non-attentional explanations for the data have been ruled out (see Spence and Driver, 1999; Ward, McDonald, and Golestani, 1998), suggesting that unimodal spatial cues may be sufficient in and of themselves to orient a person's spatial attention when presented at a clearly suprathreshold level (though see Santangelo, Belardinelli, and Spence, 2007). However, it is important to note that the behavioral evidence has always been obtained under laboratory-based research conditions in which the participants were only presented with a relatively straightforward single task setting (and in an environment that was otherwise free of distractions), which contrasts with the majority of real-world situations in which a person's attention is normally engaged by multiple competing multisensory stimuli. As a result, we recently conducted a series of experiments (Ho, Santangelo, and Spence, submitted) that were designed to examine the effectiveness of task-irrelevant spatial unimodal auditory and tactile and bimodal audiotactile cues in capturing a person's spatial attention from an otherwise highly perceptually-demanding task.

Spatial rule for effective audiotactile cuing

The experiment consisted of two conditions – no-load and high-load – presented in separate blocks of experimental trials, with the order of presentation of the two blocks counterbalanced across participants. In the no-load condition, the participants had to maintain their fixation on the center of the screen while discriminating the elevation of peripheral target circles presented equiprobably in each of the four corners of the computer screen (see Figure 7.1). A cue (one of five possible cue types) was randomly presented 240 ms prior to the presentation of the target circle. The different cue types tested included: (i) unimodal spatial white noise bursts presented on the left or right, (ii) spatial car horn sound presented on the left or right or (iii) central tactile vibration presented to the middle of the participant's stomach, or else (iv) bimodal spatial white noise bursts presented on the left or right paired with central tactile vibration or (v) bimodal spatial car horn sound presented on the left or right paired with central tactile vibration (see Table 7.1). The spatial cues were either presented on the same side as (i.e., validly-cued) or on the opposite side to (i.e., invalidly-cued) the peripheral spatial targets demanding an elevation response.

In the high-load condition, the participants had to perform the RSVP task described above in addition to the peripheral target discrimination task presented in the no-load condition. Note that on any given trial, either a target circle or a RSVP target digit, but not both, was presented, in the ratio of 3:7. The participants normally kept

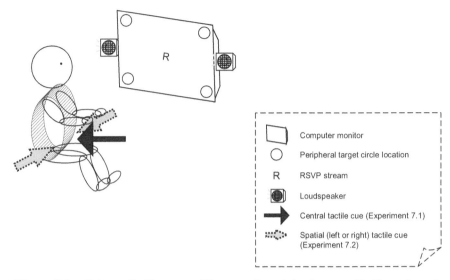

Figure 7.1   **Schematic diagram of the experimental set-up used in Experiments 7.1 and 7.2**

Table 7.1   **Summary list of the experimental conditions presented in Experiments 7.1 and 7.2**

Experiment 7.1	Cue type		
Unimodal	Spatial white noise burst		
	Spatial car horn sound		
	Central tactile vibration		
Bimodal	Left white noise burst	+	Central tactile vibration
	Right white noise burst	+	Central tactile vibration
	Left car horn sound	+	Central tactile vibration
	Right car horn sound	+	Central tactile vibration
Experiment 7.2	**Cue type**		
Unimodal	Spatial white noise burst		
	Spatial car horn sound		
	Spatial tactile vibration		
Bimodal	Left white noise burst	+	Left tactile vibration
	Right white noise burst	+	Right tactile vibration
	Left car horn sound	+	Left tactile vibration
	Right car horn sound	+	Right tactile vibration

the three footpedals that were used to collect the participants' responses depressed. They made speeded responses by lifting their toes / heel of their left / right foot depending on the response-mapping conditions to which they had been assigned. For example, some participants were instructed to lift the toes of their left foot as soon as they detected peripheral target circles presented at the top two corners of the screen (regardless of their lateral position). They also had to lift the heel of their left foot in response to peripheral target circles presented from one of the two bottom two corners of the screen, and to lift their toes of their right foot whenever they detected a RSVP target digit in the high-load condition.

The within-participants factors in this particular experiment were load (no-load vs. high-load), cue type (white noise, car horn, white noise + central tactile, or car horn + central tactile) and cuing (validly-cued vs. invalidly-cued). The results revealed a significant three-way interaction in the reaction time data (see Figure 7.2A). In particular, in the no-load condition, significant spatial cuing effects were observed for both types of unimodal cue (white noise and car horn), but surprisingly, not following the presentation of either of the bimodal cues. In addition, no significant spatial cuing effect was observed for any of the four cue types in the high-load condition.

Therefore, in contrast to the hypothesis that audiotactile (i.e., multisensory) spatial cuing might actually capture the attention of our participants more effectively than unimodal cuing, the presentation of a central tactile cue (i.e., a cue that did not provide any relevant information concerning the possible location, i.e., left or right, of the forthcoming spatial target) actually eliminated the spatial cuing effects elicited by the presentation of the spatial auditory cues (as evidenced by the results of the no-load unimodal cuing condition), resulting in a null effect of spatial cuing on performance of bimodal cuing even under conditions where the participants did not have to perform the RSVP task. This result clearly bears comparison with the sub-additive multisensory interactions reported at the cellular level by Stein and Meredith (1993; Stein and Stanford, 2008).

These results therefore suggest that even though the audiotactile cues were, if anything, more salient (in absolute terms) than the unimodal auditory cues, they nevertheless failed to capture the spatial attention of our participants. It should be noted that our results only imply that audiotactile cues failed to direct our participants' spatial attention properly when the auditory and tactile components of the cue were spatially misaligned (cf. Fitch et al., 2007, for similar results from a study showing that drivers find it no easier to localize combined spatial audiotactile cues when compared to their ability to localize the individual tactile and/or auditory cues). Although these cues failed to direct the attention of our participants spatially, it should not be taken to imply that the presentation of such cues was necessarily completely ineffective, for they may have successfully alerted our participants and hence reduced their response latencies relative to those seen under the condition when no cue was presented. One possible explanation for these null results on spatial attentional capture may be that audiotactile ventriloquism may have occurred (see Caclin, Soto-Faraco, Kingstone, and Spence, 2002). That is, the central tactile cue may have attracted the perceived location of the auditory cue from a periphery toward the center (cf. Spence and Driver, 2000; Vroomen, Bertelson, and de Gelder, 2001).

A

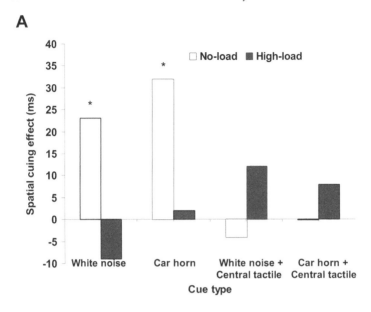

B

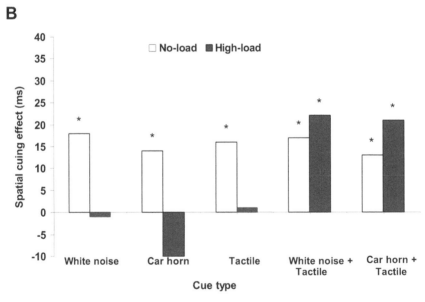

Figure 7.2   Summary graphs highlighting the mean spatial cuing effects
(ms) in the reaction time data as a function of load (no-load vs.
high-load) and cue type in Experiments 7.1 (A) and 7.2 (B). An
asterisk indicates that a significant spatial cuing effect ($p < .05$)
was observed in a particular condition

In order to determine whether directional congruency (as opposed to spatial coincidence) is a necessary prerequisite to eliciting enhanced audiotactile attentional capture, Experiment 7.2 was conducted with the auditory and tactile components of the audiotactile cues now always being presented from the congruent direction (i.e., spatial left/right white noise burst/car horn sound paired with spatial left/right tactile cue around the participant's waist). The apparatus, materials, design and procedure replicated those used in Experiment 7.1, except for the particular bimodal cue types used (see Table 7.1).

The results of Experiment 7.2 revealed a significant three-way interaction in the reaction time data (see Figure 7.2B). In contrast to the results observed in Experiment 7.1, the bimodal cues were now found to elicit significant spatial cuing effects. Specifically, in the no-load condition, all five cue types (i.e., unimodal spatial white noise, unimodal spatial car horn, unimodal spatial tactile, bimodal spatial white noise + spatial tactile, and bimodal spatial car horn + spatial tactile) showed significant spatial cuing effects in the peripheral elevation discrimination task. Critically, in the high-load condition, significant spatial cuing effects were now observed for both types of bimodal cue, but not for any of the unimodal cues.

The results of Experiment 7.2 therefore demonstrate that bimodal audiotactile cues can provide a particularly effective means of capturing the attention of an interface operator who is otherwise engaged in a highly perceptually-demanding task, when under the same conditions, unimodal cues (or warning signals) are rendered ineffective. Note that both the spatial white noise bursts (used in the majority of previous laboratory-based studies) and the sound of a car horn (a meaningful auditory cue, or icon, that is associated with the presence of another car) were shown to be equally effective in capturing our participants' attention when presented together with a spatial tactile cue. It should also be pointed out that under the no-load condition, bimodal cuing was shown to be no more effective than unimodal cuing. Importantly, however, when taken together, these results highlight the important role that spatial correspondence can play in modulating audiotactile spatial interactions.

It should be pointed out that for the case of audiovisual interaction, precise spatial coincidence of the auditory and visual components of the cue appears to be essential in order for certain facilitatory multisensory interactions to occur (e.g., Ferris, Penfold, Hameed, and Sarter, 2006). By contrast, audiotactile interactions appear to be somewhat more flexible in that they are not so constrained by the need for spatial coincidence, but rather, spatial correspondence / congruent directionality appears to be sufficient for enhanced audiotactile spatial modulation on performance to occur (cf. Kitagawa and Spence, 2006; Tajadura-Jiménez et al., submitted; Zampini, Torresan, Spence, and Murray, 2007).

Audiotactile cuing in driving

Having identified the importance of the spatial alignment of audiotactile cuing for facilitating target detection in the laboratory, we recently conducted a driving simulator study in an attempt to assess the potential effectiveness of bimodal audiotactile cuing on an interface operator's performance in the more ecologically-valid setting of the driving simulator (Ho et al., 2007b). In this study, we compared the relative effectiveness of the presentation of unimodal auditory, unimodal tactile and bimodal audiotactile warning signals in alerting and informing drivers about likely front-to-rear-end collision events. The empirical findings reported in the earlier chapters have already highlighted the potential utility of nonvisual (auditory or tactile) warning signals in improving a driver's responses to potential front-to-rear-end collisions (see also Kiefer, LeBlanc, Palmer, Salinger, Deering, and Shulman, 1999; Lee et al., 2006; Suzuki and Jansson, 2003). Given the synergistic multisensory integration effects demonstrated in Experiment 7.2, we hypothesized that the presentation of a directionally-congruent audiotactile warning signal ought to give rise to enhanced driver reactions (as reflected by faster braking responses) in response to potential collision events.

Lee et al. (2006) recently assessed the effectiveness of multisensory warning signals in terms of their ability to re-engage a distracted driver's attention during adaptive cruise control when driver intervention might be necessary. Lee et al. presented a visual icon depicting a vehicle colliding with the rear of another vehicle together with either an auditory alert tone, a seat vibration or a brake pulse, or else a combination of all four warning signals presented together. Lee et al. reported no advantage for the presentation of their combined multisensory warning signals over the other three cue types, with the fastest braking responses being observed when the auditory warning signal was presented.

The findings from the experiment reported here (Experiment 7.2) may therefore help to provide an explanation for Lee et al.'s results. Specifically, in their study, the seat vibration warning signal was not presented from a spatially-congruent direction with respect to the other warning signals, but instead the vibration came from a central position below the driver's seat. It is therefore possible that Lee et al. might have obtained performance facilitation following the presentation of their multisensory warning signals over the other cue types had their seat vibration cues been replaced with, for example, vibrations coming from the steering wheel positioned directly in front of the driver instead (see Spence and Ho, submitted).

Fifteen participants (mean age of 31 years) took part in Experiment 7.3 (Ho et al., 2007b) which was conducted at the TRL (Transport Research Laboratory) advanced driving simulator facility in Wokingham, England. The participants drove in a rural neighbourhood in a car following scenario where they were instructed to try and maintain a safe distance (1.8–2.2 s time headway) from the lead vehicle that was travelling at an average speed of 80 km/h. During the two 20-minute experimental blocks, critical events consisting of the sudden deceleration of the lead vehicle were presented at an average rate of one event per minute. A warning signal was presented concurrently with the onset of the sudden deceleration of the lead vehicle. Three types of warning signals were tested in the within-participant design: a car horn sound presented from a loudspeaker positioned on top of the instrument panel directly in front of the driver, a tactile cue presented to the front in the middle of the participant's stomach, and a combination of these two cues presented in a spatially-congruent manner (i.e., both cues were presented from the front of the driver). Both of these unimodal cues have been shown individually to be effective in our previous laboratory- and simulator-based research (see the experiments reported in the preceding three chapters).

The results of Experiment 7.3 revealed that the presentation of the audiotactile warning signals gave rise to a significant reduction in braking response latency when compared to the presentation of the tactile (19.5 per cent) and auditory (9.1 per cent) warning signals (see Figure 7.3). These results build upon the 24.7 per cent reduction in braking reaction times when tactile warning signals were presented as compared to when they were not in our earlier study (Ho et al., 2006a; see also Sklar and Sarter, 1999). Taken together, these results therefore suggest that the presentation of frontal audiotactile warning signals may offer a particularly effective means of alerting and informing drivers of a potential collision that may be about to happen on the road in front, thus leading to a performance gain (i.e., reduction in braking response latency) of as much as 40 per cent when compared to the condition when no warning signal was presented (as is the case for many cars on the road today). It should, however, be noted that the warning signals used in Experiment 7.3 were presented far more frequently than would be the case in real life. It will therefore be important in follow-up studies to ensure that the effectiveness of these warning signals is retained even when they are presented far less frequently (i.e., at a frequency that is more realistic with respect to their likely incidence in a realistic warning system; cf. Haigney and Westerman, 2001; Tufano, 1997).

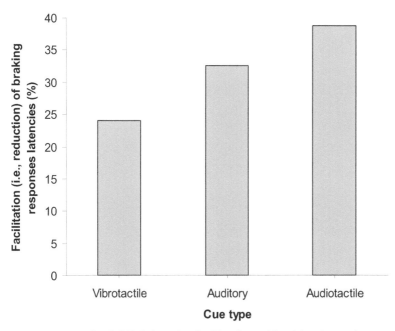

Figure 7.3   **Graph highlighting the facilitation of braking latencies reported in the simulator-based driving experiment associated with the use of vibrotactile, auditory and combined audiotactile warning signals when compared to when no warning signal was presented, as is common in many cars today (see Ho et al., 2007b)**

Taken together, the results of the three experiments reported in this chapter demonstrate the important role that spatial correspondence plays in modulating audiotactile interactions in warning signal design; in particular, its role in eliciting enhanced multisensory facilitation in the automatic disengagement and orienting of an interface operator's spatial attention. We have been able to demonstrate the importance of spatial correspondence both in the laboratory (see Experiment 7.2) and in the driving simulator (see Experiment 7.3). Our results may also provide an explanation for why multisensory warning signals might not always be effective (i.e., due to a lack of spatial correspondence; Experiment 7.1). Future experiments should therefore examine the effectiveness of these spatial multisensory warning signals under more demanding or distracting driving conditions, such as when drivers are conversing on the phone while driving, and investigate temporal parameters that may affect the design of effective multisensory warning signals (see e.g., Ferris et al., 2006). It will also be interesting in future research to investigate whether the presentation of such multisensory warning signals to car drivers may be even more effective for the growing population of older drivers (cf. Laurienti et al., 2006), since research has suggested that they may be more liable to distraction by the panoply of currently available in-car technology than are younger drivers.

Chapter 8

Conclusions

The research outlined in this book has, we hope, demonstrated the feasibility and potential effectiveness of applying the theoretical insights from the growing body of research concerning the existence of crossmodal links in spatial attention between various different sensory modalities (see Spence and Driver, 2004; Spence and Gallace, 2007) to the design of non-visual and multisensory warning signals in driving, to improve safety on the road (see Chapters 2 and 3). In particular, we have been able to show that the spatial properties inherent in certain auditory cues (either attributable to the localizable sound cues being presented spatially, or because of the informational content of verbal cues) can be used to improve the efficacy of a driver's responses to target visual driving events presented subsequently in the cued direction (see Chapter 4). In addition, facilitatory spatial cuing effects were extended from the use of auditory cues to the use of tactile cues (see Chapter 5). On the basis of the robust laboratory-based evidence regarding the beneficial effects of spatial cuing in driving, a driving simulator study was conducted that further demonstrated the significant improvement in drivers' reactions to front-to-rear-end collisions with the aid of vibrotactile cues that indicated the likely direction of the potential dangerous driving events (cf. McEvoy et al., 2007a). In particular, the performance benefit recorded in the driving simulator (a 400 ms reduction in braking latencies when compared to the no warning condition that is typical in the majority of cars today) was much larger than that reported in our laboratory-based driving studies.

Taking the research one step further, the results of the orthogonal cuing experiments reported in Chapter 6 showed that the mental processing of information by drivers during a potential collision event can be improved both at the perceptual and decisional stages, and consequently facilitate the drivers' behavioural responses (i.e., collision avoidance in the experiments reported here). To this end, the research to date demonstrates that attentional facilitation (or perceptual enhancement) will only occur if the warning signal and target event are both located within the same functional region of space, while decisional facilitation (or response priming) can occur if the cue and target are directionally congruent (even if the two stimuli occur in different functional regions of space). These findings imply that response compatibility may be an effective tool in the design of multisensory interfaces (Proctor et al., 2005), with additional benefits attributable to attentional facilitation evidenced only if interface designers consider rigorously the spatial distribution of the relevant events (i.e., of the spatial relationship between the critical driving events and the cues used to warn the driver about these events; Ho et al., submitted).

Distinct functional regions of space

Indeed, recent cognitive neuroscience research has shown that the stimuli occurring in peripersonal space are treated differently by the brain than those stimuli that are presented in extrapersonal space (see Figure 8.1; Graziano, Gross, Taylor, and Moore, 2004; Graziano, Reiss, and Gross, 1999; Kitagawa and Spence, 2006; Previc, 1998, 2000; Rizzolatti et al., 1997; Weiss et al., 2000). For instance, stimuli in near-peripersonal space may be considered to be more behaviourally relevant, thus demanding more immediate attention. As seen in Chapter 6, the attentional facilitation (and/or multisensory integration) effects may be maximal only when the cue and target events are actually located in the same functional region of space in a directionally congruent manner. This notion of cue-target co-location is consistent with those models of multisensory integration that predict maximal integration for stimuli originating from the same spatial location at approximately the same time (cf. Colonius and Diederich, 2004; Stein and Meredith, 1993; Stein and Stanford, 2008; though see Murray et al., 2005).

It is interesting to note that according to the research that has been conducted to date, there is no clearly defined borderline separating peripersonal from extrapersonal space (see Previc, 2000). Given studies showing that this boundary can be extended (i.e., it occurs at a further distance from an individual) when a person uses a tool, it is of particular interest to investigate how and why tool-use modifies body schema and spatial representations (see Holmes and Spence, 2006, for a recent review). It

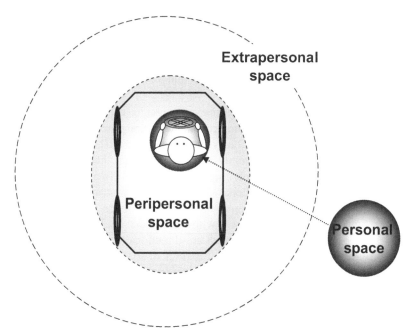

Figure 8.1   Schematic diagram of the concept of the different functional regions of space around the driver

remains an outstanding question for future research as to whether a driver in some sense 'internalizes' the car so that it temporarily becomes part of the body image (see Blakeslee and Blakeslee, 2007; Paillard, 1993, p. 40; Riddoch, 1941, p. 221). This is a particularly interesting question given that drivers' perceptions of the various regions of space around their body will determine their perceptions of the 'field of safe travel' (see Gibson and Crooks, 1938), which may ultimately relate to the issue of traffic accidents (see Snyder, Nelson, Burdick, and MacIver, 2007). Note that the region of protective space surrounding the body has been shown to extend further in the direction of sight than in other directions (e.g., Dosey and Meisels, 1969; Hall, 1966; Horowitz, Duff, and Stratton, 1964; Sommer, 1959), and that it expands somewhat when an individual feels threatened (e.g., Dosey and Meisels, 1969; Felipe and Sommer, 1966). Once again, these are clearly very interesting directions for future research, in particular in the domain of collision avoidance.

The synchronization of warning signals

In a research article published in 1999, Spence and Driver argued that the multisensory warning signals that have been used to date may not have been optimized for the human perceptual system in terms of the relative time of onset of the various sensory components of the warning signal. In particular, laboratory-based research has suggested that humans may be optimized to respond to audiovisual multisensory events occurring at a distance of approximately 10 m (e.g., Spence and Squire, 2003). At this distance, the delays in the transmission of auditory signals (i.e., sound waves) through the air cancel out the visual delays associated with the chemical transduction of light at the retina, and so both signals are thought to 'arrive' centrally in the brain at the same moment (e.g., Pöppel, 1988). There are therefore reasons to believe that humans may respond more efficiently to audiovisual warning signals presented from close to an interface operator if they are presented asynchronously such that the auditory signal is delayed somewhat with respect to the visual signal (i.e., to simulate the optimal arrival time of an event actually presented at a distance of 10 m; cf. Chan and Chan, 2006; King and Palmer, 1985). A similar argument can also be made with respect to the presentation of bimodal visuotactile warnings, given that variable delays in the arrival time of vibrotactile stimuli (when compared to visual stimuli) are associated with their transmission through the human body (see Harrar and Harris, 2005), depending upon the distance of the site of stimulation from the person's brain (Bergenheim et al., 1996; von Békésy, 1963).

The counterintuitive prediction here, based on the available cognitive neuroscience evidence, is that slightly desynchronizing the unisensory components of a multisensory warning signal (so that they will arrive at the appropriate regions of the brain in temporal synchrony) may, in fact, allow a driver (or other interface operator for that matter) to respond to that warning signal more rapidly than if both components of the warning signal had actually been presented at exactly the same time. This research topic, regarding the impact of temporal variations on the onset of the various unisensory components of multisensory warning signals clearly represents another very important subject for future research.

Taken together, therefore, the results of these previous studies suggest that while unimodal warning signals presented close to a driver's head or body may be more effective in alerting the driver, or in generating a near-automatic response from the driver, these unimodal warning signals do not necessarily lead to any attentional facilitation of the processing of events occurring in the spatial direction indicated by the signals (see Santangelo et al., 2007). Given these results, the combined use of signals presented close to the head and/or body of a driver in order to alert or arouse him or her, together with signals presented spatially (either virtually or actually) to enhance subsequent processing of any events occurring in that region of space (cf. Lee et al., 1999; Menning, Ackermann, Hertrich, and Mathiak, 2005; Previc, 2000) may, we believe, prove to represent a particularly effective form of multidimensional (or multisensory) warning signalling. Specifically, the results of the experiments reported in Chapters 6 and 7 illustrated that multisensory stimuli may give rise to enhanced performance when the components of the multisensory stimuli are presented from the same direction.

Risk compensation

A pertinent issue here relates to the automaticity of a driver's responses to the presentation of a warning signal and their increased reliance on technology. The generation of 'near-automatic' responses from a driver may be desirable under conditions where the warning signals are highly reliable and the driver is under high workload, thus having limited resources left to analyse the situation prior to producing a speeded response. However, this is not necessarily beneficial and may have adverse effects particularly if risk perception is undermined. This links to the more general topic of risk perception – the question of whether drivers really drive more safely when safety warnings are provided. In an analysis of driver behaviours after the mandatory implementation of safety features such as seat-belts and air bags in cars, researchers have found that drivers of cars equipped with such safety features tend to behave more aggressively, pushing their cars closer to the limit (that is, they exhibit a form of risk compensation; see Evans, Wasielewski, and von Buseck, 1982; Wilde, 1982). As a result, it would appear that drivers offset the original positive intention associated with safety innovation and impose additional danger or risk to their road safety instead (see Evans and Graham, 1991; Peltzman, 1975; Peterson, Hoffer, and Millner, 1995; see also Adams, 1995; Lee, forthcoming; White, Eiser, and Harris, 2004). Recent research has even reported an adverse effect of new technology, in particular automation in cars, in terms of reducing drivers' capacity to respond to emergency situations (Stanton and Young, 2005; Young and Stanton, 2004). One also needs to consider not only the impact of these new technologies on the safety of the driver, but also on the safety of other road users and pedestrians as well. However, it should also be noted that other research by Ben-Yaacov, Maltz, and Shinar (2002) has shown that drivers exposed to a rear-end collision avoidance warning system adopted longer headways for at least six months after being exposed to the safety system for a brief period.

When peripersonal space is seen via a mirror

The design of a number of the driving experiments reported in this book was based on the well-learnt association between the rear space behind a driver and its representation in the rearview mirror positioned in front of a driver. The process of learning the 'affordance' of mirrors has been argued to play an important role in shaping people's perception of the space that can only be (or, is typically) seen by means of a mirror's reflection (see Loveland, 1986). Interestingly, Binkofski et al. (2003) found that individuals with mirror apraxia (resulting from lesions of the posterior parietal cortex, that caused deficits in reaching for objects that are only visible via a mirror) can exhibit a dissociation between their body schema and their peripersonal space, with mirror apraxia affecting the processing of stimuli presented in the peripersonal space inspected via a mirror yet keeping the processing of stimuli presented on their body surface viewed via a mirror intact. Binkofski and his colleagues claimed that their findings supported the idea of a unique body schema representation that exists in the mirror space (independent of its representation in the actual world under conditions of direct viewing). These findings suggest a special processing mechanism for the representation of mirrored stimuli (see also von Fieandt, 1966).

Ecological validity

One of the major debates that have run through much of the driving research in recent years relates to the ecological validity of laboratory-based studies and, in particular, their generalizability to situations of real-world driving (see Haigney and Westerman, 2001). It would certainly be very useful therefore in future research to investigate the extent to which the findings from the laboratory-based experiments reported here scale up to modulate driving performance under more ecologically valid driving conditions. In this regard, the successful replication of the laboratory-based vibrotactile cuing research findings with the use of a high-fidelity driving simulator demonstrates the feasibility of implementing vibrotactile warning signals in real-world applications (see Chapters 5 and 7). What is even more promising is the fact that the facilitation of driving performance attributable to the presence of the vibrotactile cues was found to be much larger than might have been expected on the basis of the results of laboratory-based driving studies. The facilitatory effect in braking reaction times of a magnitude of over 1/3 s found in Experiment 5.3 is clearly important for collision avoidance considering the overall reaction time of 0.7 s and 1.5 s respectively for expected and unexpected emergency braking (Green, 2000; cf. Suetomi and Kido, 1997).

However, as noted in several of the earlier chapters, one aspect of the experimental design of all of the experiments reported in this volume that is certainly not particularly ecologically valid relates to the fact that the various warning signals that we have been assessing were always presented far more frequently than would ever be the case during real-world driving, because of our need to collect sufficient experimental data for meaningful statistical analysis.

This is an especially salient point as the front-to-rear-end collision situation modelled in many of the experiments that have been reported here is only likely to happen to a driver once every 25 years or so, on average. Even though the collision avoidance warning signals will be triggered on many more occasions than this (see Parasuraman et al., 1997), they are still far, far less frequent than the rate at which the warning signals were delivered in the experiments reported here. Clearly, then, one important task for future research will be to examine in a driving simulator (or in an on-road study) the most effective of the unisensory and multisensory warning signals uncovered by our research when presented at incidence that somewhat more closely approximates that of their predicted incidence of occurrence in real-life (or at least, with a frequency that is far lower than we were able to achieve in our own research; cf. Tufano, 1997).

Given the simple stimulation used in the present research, it is important to examine whether the present findings still hold true under conditions where more complicated patterns of vibrations are utilized and with those vibrations signalling not just collision events (front/back) but extending to signalling, for example, lane changing and/or navigational messages (e.g., van Erp and van Veen, 2004). It is therefore important to assess the quantity and quality of information that can be transmitted by the skin (see Spence and Gallace, 2007). Future studies should also examine auditory spatial cuing in a driving simulator under more realistic conditions, such as in the presence of background noise (e.g., as when a driver listens to the radio; Dibben and Williamson, 2007; Ho et al., 2007b; Slawinski and MacNeil, 2002), and with a much reduced frequency of occurrence in order to better approximate the expected rare occurrence of collision events in actual driving situations.

One may, of course, question even the ecological validity of findings from studies conducted in a driving simulator, which, to some extent, once again represent a controlled experimental environment with different demands (e.g., regarding the perception of risk and safety) from those found in real-world driving. Critics of simulator studies are typically concerned that such differences in demand characteristics may result in different behaviours on the part of drivers (e.g., Alm and Nilsson, 1995; see also Dressel and Atchley, forthcoming). However, it should be noted that several studies have now been published that have argued that driving simulator studies are both more ethical and more cost effective than on-road testing (see Haigney and Westerman, 2001; see also Murray et al., 2001), and at the same time provide good validity for measures of performance such as speed control (see Reed and Green, 1999). Although the applicability of driving simulator studies to real-world driving should of course never be asserted unquestioningly, it would seem as though simulated driving nevertheless offers an acceptable test environment for the better understanding of driver behaviour (cf. Kemeny and Panerai, 2003; McLane and Wierwille, 1975; see also Chapter 2). It is, however, worth noting that the lack of physical consequences of 'crashing' in the laboratory, or in the driving simulator, means that such studies may fail to miss the compensatory behaviours that many drivers may engage in when actually driving on the roads (see Dressel and Atchley, forthcoming; Horrey and Wickens, 2006; Reed and Green, 1999; Treffner and Barrett, 2004).

Olfactory warning signals

Another class of warning signal that may potentially be used in driving is that of olfactory signals (e.g., see Deatherage, 1972; Grayhem, Esgro, Sears, and Raudenbush, 2005; RAC Foundation, 2005). While it is not uncommon for drivers to install air fresheners to mask any malodours in their cars, some researchers have started to investigate the possibility of presenting pleasant scents in cars to calm aggressive drivers and/or to prevent them from falling asleep at the wheel. There has even been talk of introducing olfactory entertainment into the cars of the future that are capable of pumping out smells that match (i.e., are congruent with) the scenery through which the driver happens to be passing: for example, the smell of pine when driving in the forest, or the smell of fresh cut grass when driving through fields (e.g., Martin and Cooper, 2007). The ultimate aim here is to create the most multisensorially-congruent and stimulating environment for the driver of the future. To date, however, there has been a certain amount of controversy concerning the specific effects of olfactory stimulation on task performance, with a number of studies showing null effects while others have shown significant effects.

Several studies have specifically investigated the physiological effects of olfactory stimulation in various task settings (e.g., Badia, Wesensten, Lammers, Culpepper, and Harsh, 1990; Miltner, Matjak, Braun, Diekmann, and Brody, 1994; see also Spence, 2002). For example, Bensafi, Rouby, Farget, Bertrand, Vigouroux, and Holley (2002) observed a significant increase in their participants' heart rate when they were instructed to perform an odor inhalation task or a pleasantness judgment task (when performance for unpleasant odors was compared to air), but not when they had to perform a familiarity evaluation task instead. Similarly, Michael, Jacquot, Millot, and Brand (2003) have reported that ambient odors can provide a means of altering the biological significance (i.e., emotional salience) of visual stimuli, in a way that keeps the perceptual salience of the visual stimuli unchanged (see also Demattè, Österbauer, and Spence, 2007; though see also Li, Moallem, Paller, and Gottfried, 2007).

Meanwhile, other researchers have reported that the presentation of odours such as lavender, lemon, muguet or peppermint, can modulate human performance (e.g., Millot, Brand, and Morand, 2002; Raudenbush, Corley, and Eppich, 2001), cognitive processes (e.g., Baron and Thomley, 1994; Moss, Cook, Wesnes, and Duckett, 2003), and even their emotional state (Ludvigson and Rottman, 1989; Michael et al., 2003). For example, Millot et al. (2002) reported that the participants in one study responded nearly 10 per cent more rapidly to simple auditory and visual target stimuli when they were presented in the presence of an ambient odour (no matter whether it was the pleasant odour of lavender or the unpleasant odour of pyridine that smells like rotten fish) than when no odour was presented. Meanwhile, Raudenbush et al. (2001) have reported that the presence of peppermint odour can enhance athletic performance in areas such as running speed and hand-grip strength, but has no observable effect on skill-related tasks such as basketball free-throwing. Marked improvements in cognitive performance resulting from the use of ambient odour were reported in an influential study by Warm, Dember, and Parasuraman (1991). In their between-participants design, the researchers found that the periodic presentation of peppermint or muguet (lily of the valley) odour for 30 seconds every 5 minutes enabled their participants to sustain their attention for longer. That is, an overall performance improvement of approximately 20 per cent on a visual line length monitoring task was reported, as compared to the performance of a control group of participants in a no-odour (i.e., air) condition.

In order to address the limitations associated with some of the previous research, Ho and Spence (2005b) conducted an experiment designed to investigate the effect of the presentation of various odours on dual-task performance, given the mixed results highlighted in previous studies. The participants in these experiments had to detect target digits embedded amongst a stream of distractor letters, while simultaneously discriminating vibrotactile stimuli presented to the front or back of their torso that required either a compatible or incompatible foot response (see Fitts and Seeger, 1953). By manipulating the response compatibility in this manner (see Proctor et al., 2005, for a recent review of the literature on response compatibility effects in an applied context), the difficulty of the task could be varied while keeping the stimuli and task constant. To date, no published study has attempted to examine the differential effects of odours on tasks of different difficulty levels (or cognitive loads). The basic concept of the inverted U-shaped arousal function for performance suggests that people's initial level of arousal or stress may determine the resultant positive or negative interaction with the alerting or sedative stimuli presented (e.g., Degel and Koster, 1999; cf. Yerkes and Dodson, 1908; Figure 3.1).

Consistent with Warm et al.'s (1991) previous findings, the results of Ho and Spence's (2005b) study revealed that the presentation of an olfactory stimulus can facilitate dual-task performance. In particular, the participants in their study made fewer errors in the incompatible response mapping condition (i.e., responding to a frontal vibrotactile stimulus by lifting their heel, and responding to a back vibrotactile stimulus by lifting their toes) in the presence of peppermint odour than in the air (control) condition. In contrast to the general view that peppermint odour has an alerting effect (i.e., that it results in faster but less accurate performance by participants in the presence of the stimulus than in its absence; e.g., Posner, 1978), Ho and Spence argued that the presentation of the peppermint odour may have facilitated performance specifically in the incompatible response mapping condition by improving their participants' concentration. This account is based on the assumption that participants had to disrupt the natural response association by using cognitive control in the incompatible response mapping condition (cf. Austen, Soto-Faraco, Enns, and Kingstone, 2004; Rogers and Monsell, 1995).

In addition, the olfactory modulation of performance illustrated here suggests that the mixed results obtained in previous studies may be contingent upon the difficulty of the task (see e.g., Millot et al., 2002; Warm et al., 1991, for the demonstration of a significant olfactory facilitatory effect, and e.g., Gilbert, Knasko, and Sabini, 1997; Ilmberger, Heuberger, Mahrhofer, Dessovic, Kowarik, and Buchbauer, 2001, for a lack of any effect of olfactory stimulation on human performance). The peppermint odour and possibly other 'alerting' odours may only facilitate performance under more demanding experimental conditions (e.g., in the incompatible trials) but not when the tasks are relatively easy (e.g., in the compatible trials; cf. Spence et al.,

2000b, for evidence that crossmodal effects may be more pronounced under more demanding task conditions).

All in all, even though researchers are still unsure of the specific effects of certain odorants on human information processing, the potential use of odours in multisensory interface design nevertheless represents a rapidly-growing area of research interest (e.g., Bodnar, Corbett, and Nekrasovski, 2004; Bounds, 1996; Deatherage, 1972; Kaye, 2004). For instance, some researchers in the car industry have already started to investigate the possibility of stimulating this relatively underutilized sense in order to generate a more pleasurable driving experience (see Spence, 2002, for a review), to calm aggressive drivers with pleasant odours (Rinspeed, 2005; Ury, Perkins, and Goldsmith, 1972) or even to prevent drowsy drivers from falling asleep at the wheel (e.g., Baron and Kalsher, 1998; cf. Otmani, Pebayle, Roge, and Muzet, 2005; though see Badia et al., 1990; Carskadon and Herz, 2004). Of relevance here is an early study by Symons (1963) that showed improved visual sensitivity following the presentation of olfactory stimuli. It will be interesting to study the time-locked effects of odour presentation, which may be useful for the design of future warning systems, such as the design of automated presentation of in-car scent to reduce sleep-related accidents by preventing drowsy drivers from falling asleep, while avoiding the aversive elements present in many other alerting signals (e.g., Bounds, 1996; cf. McKeown and Isherwood, 2007; Parasuraman et al., 1997). The use of a simulated driving task, such as that reported in the preceding chapters, might provide a particularly appropriate means to address this question.

It is important to note, however, that odours have been shown to be minimally effective once people are asleep (e.g., Badia et al., 1990; Carskadon and Herz, 2004), and so further research on the differential effects of various olfactory stimuli on human performance at the border between wakefulness and sleep will be needed before their widespread use in cars could be recommended.

Future directions for multisensory interface design research

It is our contention that it is only by understanding the underlying nature of multisensory information processing and the brain's representation of crossmodal space (as increasingly being revealed by research in the cognitive neurosciences; see Calvert et al., 2004; Spence and Driver, 2004; Spence and Gallace, 2007) that researchers will be able to start designing multisensory warning signals that are genuinely optimized for the vagaries and information processing limitations of the human operator. One topic that is currently particularly relevant to the issue of human limitations is related to the potential problems associated with aging drivers. Older drivers are likely to suffer even more from sensory overload than younger drivers (see Dukic et al., 2006; Ponds, Brouwer, and van Wolffelaar, 1988; Verwey, 2000), yet they constitute a growing segment of the population (US Senate Special Committee on Aging, 1985–1986). It is thus interesting to investigate the possibility of supporting their sensory perceptual capacities by means of the presentation of redundant warning signals to two or more sensory modalities, or by the presentation of amplified warning signals to a given sensory modality (see Chapter 2; Spence and Ho, 2008). For instance, one recent study by Laurienti et al. (2006) reported that the elderly participants (mean age of 71 years; ranging from 65–90 years) were able to respond to simultaneously presented multisensory (auditory and visual) targets as rapidly as a group of younger participants (mean age of 28 years; ranging from 18–38 years) could respond to either of the unimodal targets. Such results are particularly promising for the design of assistive applications that support safe driving as they suggest that stimulating multiple sensory channels may represent a particularly effective compensatory strategy to help overcome the sensory decline that will increasingly be experienced by the growing population of elderly drivers.

Moreover, designers need to try to create the most ecological warning signals that can elicit an intuitive response from drivers to the appropriate spatial location where a critical emergency event may occur. The investigation into crossmodal interactions in spatial attention is suggestive of the generic spatial property in the human information processing system. Given that humans have only limited cognitive resources (e.g., Lavie, 2005; Wickens, 1984, 1992, 2002), the finding that information presented spatially in one sensory modality can enhance subsequent responses to information presented in a relevant direction or location in a different sensory modality may have important implications for multisensory interface design, particularly for situations of high informational load (cf. Bliss and Dunn, 2000), as is often the case for driving.

Further, as O'Regan et al. (1999) have pointed out, dangerous events may occur without being noticed if these events happen to coincide temporally with other harmless disturbances, such as, for example, small stones hitting on the car windscreen (see also Batchelder, Rizzo, Vanderleest, and Vecera, 2003; Simons and Rensink, 2005; Velichkovsky et al., 2002). Given that such time-locked information processing deficits also occur crossmodally (see Colavita, 1974; Sinnett, Spence, and Soto-Faraco, 2007, 2008), it is important for car manufacturers to design and install multisensory warning devices that can facilitate the appropriate deployment of a driver's attention. Finally, the insights gained from the specific study of warning signals based on cognitive neuroscience principles delineated here should not be taken to be restricted to the design of in-car warning signals only. The application of our understanding of brain mechanisms may also be extended to the design of warning signals and interfaces in a variety of other application domains, from in the air cockpits to hospital warning displays to nuclear power stations (see Oviatt, 1999, 2002). The development and implementation of multisensory interfaces that are designed around the mechanisms of our brain offer a promising future.

References

Adams, J. (1995). *Risk.* London: UCL Press.

Akamatsu, M., MacKenzie, I. S., and Hasbroucq, T. (1995). A comparison of tactile, auditory, and visual feedback in a pointing task using a mouse-type device. *Ergonomics, 38,* 816–827.

Alais, D., Morrone, C., and Burr, D. (2006). Separate attentional resources for vision and audition. *Proceedings of the Royal Society B: Biological Sciences, 273,* 1339–1345.

Alibali, M., Heath, D., and Myers, H. (2001). Effects of visibility between speaker and listener on gesture production: Some gestures are meant to be seen. *Journal of Memory and Language, 44,* 169–188.

Allport, A., and Glenn, W. (2000). Task switching and the measurement of "switch costs". *Psychological Research, 63,* 212–233.

Allport, A., and Hsieh, S. (2001). Task-switching: Using RSVP methods to study an experimenter-cued shift of set. In K. Shapiro (Ed.), *The limits of attention: Temporal constraints in human information processing* (pp. 36–64). Oxford: Oxford University Press.

Alm, H., and Nilsson, L. (1994). Changes in driver behaviour as a function of hands free mobile phones – A simulator study. *Accident Analysis and Prevention, 26,* 441–451.

Alm, H., and Nilsson, L. (1995). The effects of a mobile telephone task on driver behaviour in a car following situation. *Accident Analysis and Prevention, 27,* 707–715.

Alm, H., and Nilsson, L. (2001). The use of car phones and changes in driver behaviour. *International Journal of Vehicle Design, 26,* 4–11.

Amado, S., and Ulupinar, P. (2005). The effects of conversation on attention and peripheral detection: Is talking with a passenger and talking on the cell phone different? *Transportation Research Part F, 8,* 383–395.

Arent, S. M., and Landers, D. M. (2003). Arousal, anxiety, and performance: A reexamination of the inverted-U hypothesis. *Research Quarterly for Exercise and Sport, 74,* 436–444.

Arthur, W., Jr., Barrett, G. V., and Alexander, R. A. (1991). Prediction of vehicular accident involvement: A meta-analysis. *Human Performance, 4,* 89–105.

Arthur, W. Jr., and Doverspike, D. (1992). Locus of control and auditory selective attention as predictors of driving accident involvement: A comparative longitudinal investigation. *Journal of Safety Research, 23,* 73–80.

Arthur, W., and Strong, M. H. (1994). Validation of visual attention test as a predictor of driving accident involvement. *Journal of Occupational and Organizational Psychology, 67,* 173–183.

Arthur, W., Strong, M. H., and Williamson, J. (1994). Validation of visual attention test as a predictor of driving accident involvement. *Journal of Occupational and Organizational Psychology, 67,* 173–182.

Ashley, S. (2001). Driving the info highway. *Scientific American, 285(4),* 44–50.

Austen, E. L., Soto-Faraco, S., Enns, J. T., and Kingstone, A. (2004). Mislocalizations of touch to a fake hand. *Cognitive, Affective, and Behavioral Neuroscience, 4,* 170–181.

Auvray, M., Gallace, A., Tan, H. Z., and Spence, C. (2007). Crossmodal change blindness between vision and touch. *Acta Psychologica, 126,* 79–97.

Avolio, B. J., Kroeck, K. G., and Panek, P. E. (1985). Individual differences in information-processing ability as a predictor of motor vehicle accidents. *Human Factors, 27,* 577–588.

Bach-y-Rita, P. (2004). Tactile sensory substitution studies. *Annals of the New York Academy of Sciences, 1013,* 83–91.

Badia, P., Wesensten, N., Lammers, W., Culpepper, J., and Harsh, J. (1990). Responsiveness to olfactory stimuli presented in sleep. *Physiology and Behavior, 48,* 87–90.

Barkana, Y., Zadok, D., Morad, Y., and Avni, I. (2004). Visual field attention is reduced by concomitant hands-free conversation on a cellular telephone. *American Journal of Ophthalmology, 138,* 347–353.

Baron, R. A., and Kalsher, M. J. (1998). Effects of a pleasant ambient fragrance on simulated driving performance: The sweet smell of... safety? *Environment and Behavior, 30,* 535–552.

Baron, R. A., and Thomley, J. (1994). A whiff of reality: Positive affect as a potential mediator of the effects of pleasant fragrances on task performance and helping. *Environment and Behavior, 26,* 766–784.

Batchelder, S., Rizzo, M., Vanderleest, R., and Vecera, S. (2003). Traffic scene related change blindness in older drivers. *Proceedings of the 2nd International Driving Symposium on Human Factors in Driver Assessment, Training, and Vehicle Design,* 177–181.

Beede, K. E., and Kaas, S. J. (2006). Engrossed in conversation: The impact of cell phones on simulated driving performance. *Accident Analysis and Prevention, 38,* 415–421.

Beck, D. M., Rees, G., Frith, C. D., and Lavie, N. (2001). Neural correlates of change detection and change blindness. *Nature Neuroscience, 4,* 645–650.

Begault, D. R. (1993). Head-up auditory displays for traffic collision avoidance system advisories: A preliminary investigation. *Human Factors, 35,* 707–717.

Begault, D. R. (1994). *3-D sound for virtual reality and multimedia.* Cambridge, MA: Academic Press.

Beh, H. C., and Hirst, R. (1999). Performance on driving-related tasks during music. *Ergonomics, 42,* 1087–1098.

Belz, S. M., Robinson, G. S., and Casali, J. G. (1999). A new class of auditory warning signals for complex systems: Auditory icons. *Human Factors, 41,* 608–618.

Ben-Yaacov, A., Maltz, M., and Shinar, D (2002). Effects of an in-vehicle collision avoidance warning system on short- and long-term driving performance. *Human Factors, 44,* 335–342.

Bensafi, M., Rouby, C., Farget, V., Bertrand, B., Vigouroux, M., and Holley, A. (2002). Autonomic nervous system responses to odours: The role of pleasantness and arousal. *Chemical Senses, 27,* 703–709.

Bergenheim, M., Johansson, H., Granlund, B., and Pedersen, J. (1996). Experimental evidence for a sensory synchronization of sensory information to conscious experience. In S. R. Hameroff, A. W. Kaszniak, and A. S. Scott (Eds.), *Toward a science of consciousness: The first Tucson discussions and debates* (pp. 303–310). Cambridge, MA: MIT Press.

Berger, A., Henik, A., and Rafal, R. (2005). Competition between endogenous and exogenous orienting of visual attention. *Journal of Experimental Psychology: General, 134*, 207–221.

Bhargava, A., Scott, M., Traylor, R., Chung, R., Mrozek, K., Wolter, J., et al. (2005). Effect of cognitive load on tactor location identification in zero-g. *Proceedings of the 2005 World Haptics Conference*, 56–62.

Binkofski, F., Buccino, G., Dohle, C., Seitz, R. J., and Freund, H. J. (1999). Mirror agnosia and mirror ataxia constitute different parietal lobe disorders. *Annals of Neurology, 46*, 51–61.

Binkofski, F., Butler, A., Buccino, G., Heide, W., Fink, G., Freund, H.-J., et al. (2003). Mirror apraxia affects the peripersonal mirror space. A combined lesion and cerebral activation study. *Experimental Brain Research, 153*, 210–219.

Blakeslee, S., and Blakeslee, M. (2007). *The body has a mind of its own: How body maps in your brain help you do (almost) anything better*. New York: Random House.

Blattner, M. M., Sumikawa, D. A., and Greenberg, R. M. (1989). Earcons and icons: Their structure and common design principles. *Human-Computer Interaction, 4*, 11–44.

Bliss, J. P. (2003). An investigation of alarm related incidents in aviation. *International Journal of Aviation Psychology, 13*, 249–268.

Bliss, J. P., and Acton, S. A. (2003). Alarm mistrust in automobiles: How collision alarm reliability affects driving. *Applied Ergonomics, 34*, 499–509.

Bliss, J. P., and Dunn, M. C. (2000). Behavioural implications of alarm mistrust as a function of task workload. *Ergonomics, 43*, 1283–1300.

Boase, M., Hannigan, S., and Porter, J. M. (1988). Sorry, can't talk now ... just overtaking a lorry: The definition and experimentation investigation of the problem of driving and handsfree carphone use. *Contemporary Ergonomics: Proceedings of the Ergonomics Society's 1988 Annual Conference*, 527–532.

Bodnar, A., Corbett, R., and Nekrasovski, D. (2004). AROMA: Ambient awareness through olfaction in a messaging application. *Proceedings of the 6th International Conference on Multimodal Interfaces*, 183–190.

Boltz, M., Schulkind, M., and Kantra, S. (1991). Effects of background music on the remembering of filmed events. *Memory and Cognition, 19*, 593–606.

Booher, H. R. (1978). Effects of visual and auditory impairment in driving performance. *Human Factors, 20*, 307–320.

Bounds, W. (1996, May 6). Sounds and scents to jolt drowsy drivers. *Wall Street Journal*, B1.

Bowditch, S. C. (2001). Driver distraction: A replication and extension of Brown, Tickner and Simmons (1969). In G. B. Grayson (Ed.), *Behavioural research in road safety 11*. Wokingham: Transport Research Laboratory.

Brewster, S., and Brown, L. M. (2004). Tactons: Structured tactile messages for non-visual information display. *Proceedings of the 5th Australasian User Interface Conference*, 15–23.

Breznitz, S. (1984). *Cry wolf: The psychology of false alarms*. Hillsdale, NJ: Erlbaum.

Briem, V., and Hedman, L. R. (1995). Behavioural effects of mobile telephone use during simulated driving. *Ergonomics, 38*, 2536–2562.

Bristow, D., Haynes, J.-D., Sylvester, R., Frith, C. D., and Rees, G. (2005). Blinking suppresses the neural response to unchanging retinal stimulation. *Current Biology, 15*, 1296–1300.

Broadbent, D. E. (1958). *Perception and communication*. Elmsford, NJ: Pergamon.

Brodsky, W. (2002). The effects of music tempo on simulated driving performance and vehicular control. *Transportation Research Part F: Traffic Psychology and Behaviour, 4*, 219–241.

Bronkhorst, A. W., Veltman, J. A., and van Breda, L. (1996). Application of a three-dimensional auditory display in flight task. *Human Factors, 38*, 23–33.

Brookhuis, K. A., de Vries, G., and de Waard, D. (1991). The effects of mobile telephoning on driving performance. *Accident Analysis and Prevention, 23*, 309–316.

Brookhuis, K., de Waard, D., and Mulder, B. (1994). Measuring driver performance by car-following in traffic. *Ergonomics, 37*, 427–434.

Brown, I. D. (1962). Measuring the 'spare mental capacity' of car drivers by a subsidiary auditory task. *Ergonomics, 5*, 247–250.

Brown, I. D. (1965). Effect of a car radio on driving in traffic. *Ergonomics, 8*, 475–479.

Brown, I. D., and Poulton, E. C. (1961). Measuring the spare 'mental capacity' of car drivers by a subsidiary task. *Ergonomics, 4*, 35–40.

Brown, I. D., Simmonds, D. C. V., and Tickner, A. H. (1967). Measurement of control skills, vigilance, and performance on a subsidiary task during 12 hours of car driving. *Ergonomics, 10*, 665–673.

Brown, I. D., Tickner, A. H., and Simmonds, D. C. (1969). Interference between concurrent tasks of driving and telephoning. *Journal of Applied Psychology, 53*, 419–424.

Brown, L. M., Brewster, S. A., and Purchase, H. C. (2005). A first investigation into the effectiveness of tactons. *Proceedings of the World Haptics Conference 2005*, 167–176.

Bruce, D., Boehm-Davis, D. A., and Mahach, K. (2000). In-vehicle auditory display of symbolic information. *Proceedings of the IEA 2000/HFES 2000 Congress*, 3-230-3-233.

Bryan, W. E. (1957). Research in vision and traffic safety. *Journal of the American Optometric Association, 29*, 169–172.

Bull, M. (2005). Soundscapes of the car: A critical study of automobile habituation. In M. Bull and L. Black (Eds.), *The auditory culture reader* (pp. 357–374). Oxford: Berg.

Burke, M. W., Gilson, R. D., and Jagacinski, R. J. (1980). Multi-modal information processing for visual workload relief. *Ergonomics, 23*, 961–975.

Burnett, G. E., and Joyner, S. M. (1997). An assessment of moving map and symbol-based route guidance systems. In Y. I. Noy (Ed.), *Ergonomics and safety of intelligent driver interfaces* (pp. 115–137). Mahwah, NJ: Lawrence Erlbaum.

Burt, J. L., Bartolome, D. S., Burdette, D. W., and Comstock, J. R., Jr. (1995). A psychophysiological evaluation of the perceived urgency of auditory warning signals. *Ergonomics, 38*, 2327–2340.

Cabrera, D., Ferguson, S., and Laing, G. (2005). Development of auditory alerts for air traffic control consoles. *Audio Engineering Society 119ʰ Convention*, 1–21.

Caclin, A., Soto-Faraco, S., Kingstone, A., and Spence, C. (2002). Tactile "capture" of audition. *Perception and Psychophysics, 64*, 616–630.

Caird, J. K., Willness, C. R., Steel, P., and Scialfa, C. (forthcoming). A meta-analysis of the effects of cell phones on driver performance. *Accident Analysis and Prevention*.

Calvert, G. A., Spence, C., and Stein, B. E. (Eds.). (2004). *The handbook of multisensory processes*. Cambridge, MA: MIT Press.

Campbell, J. L., Hooey, B. L., Camey, C., Hanowski, R. J., Gore, B. F., Kantowitz, B. H., et al. (1996). *Investigation of alternative displays for side collision avoidance systems* (Report No. DOT HS 808 579). Seattle, WA: Battelle Human Factors Transportation Center.

Carskadon, M. A., and Herz, R. S. (2004). Minimal olfactory perception during sleep: Why odor alarms will not work for humans. *Sleep, 27*, 402–405.

Cassel, E. E., and Dallenbach, K. M. (1918). The effect of auditory distraction upon the sensory reaction. *American Journal of Psychology, 29*, 129–143.

Catchpole, K. R., McKeown, J. D., and Withington, D. J. (1999a). Alerting, informing and localisable auditory warnings. In D. Harris (Ed.), *Engineering psychology and cognitive ergonomics, Vol. 4: Job design, product design and human-computer interaction* (pp. 447–454). Aldershot, England: Ashgate.

Catchpole, K. R., McKeown, J. D., and Withington, D. J. (1999b). Localisable auditory warnings: Integral 'where' and 'what' components. In D. Harris (Ed.), *Engineering psychology and cognitive ergonomics, Vol. 4: Job design, product design and human-computer interaction* (pp. 439–446). Aldershot, England: Ashgate.

Catchpole, K. R., McKeown, J. D., and Withington, D. J. (2004). Localizable auditory warning pulses. *Ergonomics, 47*, 748–771.

Chamorro-Premuzic, T., and Furnham, A. (2007). Personality and music: Can traits explain how people use music in everyday life? *British Journal of Psychology, 98*, 175–185.

Chan, A. H. S., and Chan, K. W. L. (2006). Synchronous and asynchronous presentation of auditory and visual signals: Implications for control console design. *Applied Ergonomics, 37*, 131–140.

Chan, J. S., Merrifield, K., and Spence, C. (2005). Auditory spatial attention assessed in a flanker interference task. *Acta Acustica, 91*, 554–563.

Cherry, E. C. (1953). Some experiments upon the recognition of speech with one and two ears. *Journal of the Acoustical Society of America, 25*, 975–979.

Cherry, E. C. (1954). Some further experiments upon the recognition of speech, with one and with two ears. *Journal of the Acoustical Society of America, 26*, 554–559.

Cholewiak, R. W., Brill, J. C., and Schwab, A. (2004). Vibrotactile localization on the abdomen: Effects of place and space. *Perception and Psychophysics, 66*, 970–987.

Cnossen, F., Meijman, T., and Rothengatter, T. (2004). Adaptive strategy changes as a function of task demands: A study of car drivers. *Ergonomics, 47*, 218–236.

Cockburn, A., and Brewster, S. (2005). Multimodal feedback for the acquisition of small targets. *Ergonomics, 48*, 1129–1150.

Coffman, D. D., Gfeller, K., and Eckert, M. (1995). Effect of textual setting, training, and gender on emotional responses to verbal and musical information. *Psychomusicology, 14*, 117–136.

Cohen, J. T., and Graham, J. D. (2003). A revised economic analysis of restrictions on the use of cell phones while driving. *Risk Analysis, 23*, 5–17.

Colavita, F. B. (1974). Human sensory dominance. *Perception and Psychophysics, 16*, 409–412.

Collet, C., Petit, C., Priez, A., and Dittmar, A. (2005). Stroop color-word test, arousal, electrodermal activity and performance in a critical driving situation. *Biological Psychology, 69*, 195–203.

Colonius, H., and Diederich, A. (2004). Multisensory interaction in saccadic reaction time: A time-window-of-integration model. *Journal of Cognitive Neuroscience, 16*, 1000–1009.

Consiglio, W., Driscoll, P., Witte, M., and Berg, W. P. (2003). Effect of cellular telephone conversations and other potential interference on reaction time in a braking response. *Accident Analysis and Prevention, 35*, 495–500.

Conway, A. R. A., Cowan, N., and Bunting, M. F. (2001). The cocktail party phenomenon revisited: The importance of working memory capacity. *Psychonomic Bulletin and Review, 8*, 331–335.

Cooper, P. J., Zheng, Y., Richard, C., Vavrik, J., Heinrichs, B., and Sigmund, G. (2003). The impact of hands-free message reception/response on driving task performance. *Accident Analysis and Prevention, 35*, 23–35.

Crundall, D., Bains, M., Chapman, P., and Underwood, G. (2005). Regulating conversation during driving: A problem for mobile telephones? *Transportation Research Part F: Traffic Psychology and Behaviour, 8*, 197–211.

Davenport, W. G. (1972). Vigilance and arousal: Effects of different types of background stimulation. *Journal of Psychology, 82*, 339–346.

Deatherage, B. H. (1972). Auditory and other sensory forms of information presentation. In H. P. Van Cott and R. G. Kinkade (Eds.), *Human engineering guide to equipment design* (pp. 123–160). New York: John Wiley and Sons.

de Fockert, J. W., Rees, G., Frith, C. D., and Lavie, N. (2001). The role of working memory in visual selective attention. *Science, 291*, 1803–1806.

Degel, J., and Koster, E. P. (1999). Odors: Implicit memory and performance effects. *Chemical Senses, 24*, 317–325.

Demattè, M. L., Österbauer, R., and Spence, C. (2007). Olfactory cues modulate judgments of facial attractiveness. *Chemical Senses, 32*, 603–610.

Dewar, R. E. (1988). In-vehicle information and driver overload. *International Journal of Vehicle Design, 9*, 557–564.

Dibben, N., and Williamson, V. J. (2007). An exploratory survey of in-vehicle music listening. *Psychology of Music, 35*, 571–589.

Dingus, T. A., Hulse, M. C., Mollenhauer, M. A., Fleischman, R. N., McGehee, D. V., and Manakkal, N. (1997). Effects of age, system experience, and navigation technique on driving with an advanced traveler information system. *Human Factors, 39*, 177–199.

Doherty, S. T., Andrey, J. C., and MacGregor, C. (1998). The situational risks of young drivers: The influence of passengers, time of day and day of week on accident rates. *Accident Analysis and Prevention, 30*, 45–52.

Doll, T. J., and Hanna, T. E. (1989). Enhanced detection with bimodal sonar display. *Human Factors, 31*, 539–550.

Dosey, M. A., and Meisels, M. (1969). Personal space and self-protection. *Journal of Personality and Social Psychology, 11*, 93–97.

Doverspike, D., Cellar, D., and Barrett, G. V. (1986). The auditory selective attention test: A review of field and laboratory studies. *Educational and Psychological Measurement, 46*, 1095–1103.

Dressel, J., and Atchley, P. (forthcoming). Cellular phone use while driving: A methodological checklist for investigating dual-task costs. *Transportation Research Part F.*

Drews, F. A., Pasupathi, M., and Strayer, D. L. (2004). Passenger and cell-phone conversation in simulated driving. *Proceedings of the Human Factors and Ergonomics Society 48th Annual Meeting 2004*, 2210–2212.

Driver, J. (2001). A selective review of selective attention research from the past century. *British Journal of Psychology, 92*, 53–78.

Driver, J., and Spence, C. J. (1994). Spatial synergies between auditory and visual attention. In C. Umiltà and M. Moscovitch (Eds.), *Attention and performance XV: Conscious and nonconscious information processing* (pp. 311–331). Cambridge, MA: MIT Press.

Driver, J., and Spence, C. (2004). Crossmodal spatial attention: Evidence from human performance. In C. Spence and J. Driver (Eds.), *Crossmodal space and crossmodal attention* (pp. 179–220). Oxford: Oxford University Press.

Driving Standards Agency (2004). *The highway code.* London: Department for Transport.

Drory, A. (1985). Effects of rest and secondary task on simulated truck-driving task performance. *Human Factors, 27*, 201–207.

Dukic, T., Hanson, L., and Falkmer, T. (2006). Effect of drivers' age and push button locations on visual time off road, steering wheel deviation and safety perception. *Ergonomics, 49*, 78–92.

Eby, D. W., Vivoda, J. M., and St Louis, R. M. (2006). Driver hand-held cellular phone use: A four year analysis. *Journal of Safety Research, 37*, 261–265.

Edworthy, J., and Hellier, E. (2006). Complex nonverbal auditory signals and speech warnings. In M. S. Wogalter (Ed.), *Handbook of warnings* (pp. 199–220). Mahwah, NJ: Lawrence Erlbaum.

Edworthy, J., Loxley, S., and Dennis, I. (1991). Improving auditory warning design: Relationship between warning sound parameters and perceived urgency. *Human Factors, 33*, 205–231.

Eimer, M. (1999). Can attention be directed to opposite directions in different modalities? An ERP study. *Clinical Neurophysiology, 110*, 1252–1259.

Elander, J., West, R., and French, D. (1993). Behavioral correlates of individual differences in road-traffic crash risk: An examination of methods and findings. *Psychological Bulletin, 113*, 279–294.

Enriquez, M., Afonin, O., Yager, B., and Maclean, K. (2001). A pneumatic tactile alerting system for the driving environment. *Proceedings of the 2001 Workshop on Perceptive User Interfaces*, 1–7.

Enriquez, M. J., and MacLean, K. E. (2004). Impact of haptic warning signal reliability in a time-and-safety-critical task. *Proceedings of the 12th International Symposium on Haptic Interfaces for Virtual Environment and Teleoperator Systems*, 407–414.

Enriquez, M., MacLean, K. E., and Chita, C. (2006). Haptic phonemes: Basic building blocks of haptic communication. *Proceedings of the 8th International Conference on Multimodal Interfaces*, 302–309.

Evans, L. (1991). *Traffic safety and the driver*. New York: Van Nostrand Reinhold.

Evans, L., and Wasielewski, P. (1982). Do accidents involved drivers exhibit riskier everyday driving behavior? *Accident Analysis and Prevention, 14*, 57–64.

Evans, L., Wasielewski, P., and von Buseck, C. R. (1982). Compulsory seat belt usage and driver risk-taking behavior. *Human Factors, 24*, 41–48.

Evans, W. N., and Graham, J. D. (1991). Risk reduction or risk compensation? The case of mandatory safety-belt use laws. *Journal of Risk and Uncertainty, 4*, 61–73.

Eves, D. A., and Novak, M. M. (1998). Extraction of vector information using a novel tactile display. *Displays, 18*, 169–181.

Fafrowicz, M., and Marek, T. (2007). Quo vadis, neuroergonomics? *Ergonomics, 50*, 1941–1949.

Fagerstrom, K.-O., and Lisper, H.-O. (1977). Effects of listening to car radio, experience, and personality of the driver on subsidiary reaction time and heart rate in a long-term driving task. In R. R. Mackie (Ed.), *Vigilance: Theory, operational performance, and physiological correlates* (pp. 73–85). New York: Plenum.

Fagioli, S., and Ferlazzo, F. (2006). Shifting attention across spaces while driving: Are hands-free mobile phones really safer? *Cognitive Processing, 7*, S147.

Fairclough, S. H., Ashby, M. C., and Parkes, A. M. (1993). In-vehicle displays, visual workload and usability evaluation. In A. G. Gale, I. D. Brown, C. M. Haslegrave, H. W. Kruysse, and S. P. Taylor (Eds.), *Vision in vehicles – IV* (pp. 245–254). Amsterdam: Elsevier Science.

Felipe, N. J., and Sommer, R. (1966). Invasions of personal space. *Social Problems, 14*, 206–214.

Fenton, R. E. (1966). An improved man-machine interface for the driver-vehicle system. *IEEE Transactions on Human Factors in Electronics, HFE-7*, 150–157.

Fernandez-Duque, D., and Thornton, I. M. (2000). Change detection without awareness: Do explicit reports underestimate the representation of change in the visual system? *Visual Cognition, 7*, 323–344.

Ferris, T., Penfold, R., Hameed, S., and Sarter, N. (2006). The implications of crossmodal links in attention for the design of multimodal interfaces: A driving simulator study. *Proceedings of the Human Factors and Ergonomics Society 50th Annual Meeting*, 406–409.

Fidell, S. (1982). Comments on Mulligan and Shaw's "Multimodal signal detection: Independent decisions vs. integration". *Perception and Psychophysics, 31*, 90.

Fitch, G. M., Kiefer, R. J., Hankey, J. M., and Kleiner, B. M. (2007). Toward developing an approach for alerting drivers to the direction of a crash threat. *Human Factors, 49*, 710–720.

Fitts, P. M., and Seeger, G. M. (1953). S-R compatibility: Spatial characteristics of stimulus and response codes. *Journal of Experimental Psychology, 46*, 199–210.

Fontaine, C. W., and Schwalm, N. D. (1979). Effects of familiarity of music on vigilant performance. *Perceptual and Motor Skills, 49*, 71–74.

Franconeri, S. L., Simons, D. J., and Junge, J. A. (2004). Searching for stimulus-driven shifts of attention. *Psychonomic Bulletin and Review, 11*, 876–881.

Furnham, A., and Strbac, L. (2002). Music is as distracting as noise: The differential distraction of background music and noise on the cognitive test performance of introverts and extraverts. *Ergonomics, 45*, 203–217.

Fuse, T., Matsunaga, K., Shidoji, K., Matsuki, Y., and Umezaki, K. (2001). The cause of traffic accidents when drivers use car phones and the functional requirement of car phones for safe driving. *International Journal of Vehicle Design, 26*, 48–56.

Gallace, A., Tan, H. Z., and Spence, C. (2007). The body surface as a communication system: The state of the art after 50 years. *Presence: Teleoperators and Virtual Environments, 16*, 655–676.

Garcia-Larrea, L., Perchet, C., Perrin, F., and Amenedo, E. (2001). Interference of cellular phone conversations with visuomotor tasks: An ERP study. *Journal of Psychophysiology, 15*, 14–21.

Gaver, W. W. (1986). Auditory icons: Using sound in computer interfaces. *Human-Computer Interaction, 2*, 167–177.

Gaver, W. W. (1989). The SonicFinder: An interface that uses auditory icons. *Human-Computer Interaction, 4*, 67–94.

Gaver, W. W. (1993a). How do we hear in the world? Explorations in ecological acoustics. *Ecological Psychology, 5*, 285–313.

Gaver, W. W. (1993b). What in the world do we hear? An ecological approach to auditory event perception. *Ecological Psychology, 5*, 1–29.

Gaver, W. W., Smith, R. B., and O'Shea, T. (1991). Effective sounds in complex systems: The ARKOLA simulation. *Proceedings of the SIGCHI Conference on Human Factors in Computing Systems: Reaching Through Technology*, 85–90.

Geissler, L. R. (1915). Sound localization under determined expectation. *American Journal of Psychology, 26*, 268–285.

Geldard, F. A. (1960). Some neglected possibilities of communication. *Science, 131*, 1583–1588.

Gescheider, G. A., and Niblette, R. K. (1967). Cross-modality masking for touch and hearing. *Journal of Experimental Psychology, 74*, 313–320.

Gibson, J. J., and Crooks, L. E. (1938). A theoretical field-analysis of automobile-driving. *American Journal of Psychology, 51*, 453–471.

Gilbert, A. N., Knasko, S. C., and Sabini, J. (1997). Sex differences in task performance associated with attention to ambient odor. *Archives of Environmental Health, 52*, 195–199.

Gilliland, K., and Schlegel, R. E. (1994). Tactile stimulation of the human head for information display. *Human Factors, 36*, 700–717.

Gilmer, B. v. H. (1961). Toward cutaneous electro-pulse communication. *Journal of Psychology, 52*, 211–222.

Gilson, R. D., and Fenton, R. E. (1974). Kinesthetic-tactual information presentations – Inflight studies. *IEEE Transactions on Systems, Man, and Cybernetics, SMC-4*, 531–535.

Glassbrenner, D. (2005). Driver cell phone use in 2004 – Overall results. In *Traffic safety facts: Research note* (DOT HS 809 847). Washington, DC: U. S. Department of Transportation. Retrieved February 17, 2008, from http://www-nrd.nhtsa.dot.gov/pdf/nrd-30/NCSA/RNotes/2005/809847.pdf

Godthelp, H., and Schumann, J. (1993). Intelligent accelerator: An element of driver support. In A. M. Parkes and S. Franzen (Eds.), *Driving future vehicles* (pp. 265–274). London: Taylor and Francis.

Goodman, M. J., Tijerina, L., Bents, F. D., and Wierwille, W. W. (1999). Using cellular telephones in vehicles: Safe or unsafe? *Transportation Human Factors, 1*, 3–42.

Gopher, D., and Kahneman, D. (1971). Individual differences in attention and the prediction of flight criteria. *Perceptual and Motor Skills, 33*, 1335–1342.

Gopher, D. (1982). A selective attention test as a predictor of success in flight training. *Human Factors, 24*, 173–183.

Graham, R. (1999). Use of auditory icons as emergency warnings: Evaluation within a vehicle collision avoidance application. *Ergonomics, 42*, 1233–1248.

Graham, R., and Carter, C. (2001). Voice dialling can reduce the interference between concurrent tasks of driving and phoning. *International Journal of Vehicle Design, 26*, 30–47.

Graham-Rowe, D. (2001). Asleep at the wheel. *New Scientist, 169 (2283)*, 24.

Grayhem, R., Esgro, W., Sears, T., and Raudenbush, B. (2005). *Effects of odor administration on driving performance, safety, alertness, and fatigue.* Poster presented at the 27th Annual Meeting of the Association for Chemoreception Sciences, Sarasota, FL.

Graziano, M. S. A., Gross, C. G., Taylor, C. S. R., and Moore, T. (2004). A system of multimodal areas in the primate brain. In C. Spence and J. Driver (Eds.), *Crossmodal space and crossmodal attention* (pp. 51–67). Oxford: Oxford University Press.

Graziano, M. S. A., Reiss, L. A. J., and Gross, C. G. (1999). A neuronal representation of the location of nearby sounds. *Nature, 397*, 428–430.

Green, M. (2000). "How long does it take to stop?" Methodological analysis of driver perception-brake times. *Transportation Human Factors, 2*, 195–216.

Gregory, R. (1998). *Mirrors in mind.* London: Penguin Books.

Gugerty, L., Rando, C., Rakauskas, M., Brooks, J., and Olson, H. (2003). Differences in remote versus in-person communications while performing a driving task. *Proceedings of the Human Factors and Ergonomics Society 47th Annual Meeting*, 1855–1859.

Gugerty, L., Rakauskas, M., and Brooks, J. (2004). Effects of remote and in-person verbal interactions on verbalization rates and attention to dynamic spatial scenes. *Accident Analysis and Prevention, 36*, 1029–1043.

Guillaume, A., Pellieux, V., Chastres, V., and Drake, C. (2003). Judging the urgency of nonvocal auditory warning signals: Perceptual and cognitive processes. *Journal of Experimental Psychology: Applied, 9*, 196–212.

Haas, E. C., and Casali, J. G. (1995). Perceived urgency and response time to multi-tone and frequency-modulated warning signals in broadband noise. *Ergonomics, 38*, 2313–2326.

Haas, E. C., and Edworthy, J. (1996). Designing urgency into auditory warnings using pitch, speed and loudness. *Computing and Control Engineering Journal, 7*, 193–198.

Haas, M. W. (1995). Virtually-augmented interfaces for tactical aircraft. *Biological Psychology, 40*, 229–238.

Hahn, R. W., Tetlock, P. C., and Burnett, J. K. (2000). Should you be allowed to use your cellular phone while driving? *Regulation, 23(3)*, 46–55.

Haigney, D. E., Taylor, R. G., and Westerman, S. J. (2000). Concurrent mobile (cellular) phone use and driving performance: Task demand characteristics and compensatory process. *Transportation Research Part F, 3*, 113–121.

Haigney, D., and Westerman, S. J. (2001). Mobile (cellular) phone use and driving: A critical review of research methodology. *Ergonomics, 44*, 132–143.

Hall, E. T. (1966). *The hidden dimension: Man's use of space in public and private.* London: Bodley Head.

Hancock, P. A., Lesch, M., and Simmons, L. (2003). The distraction effects of phone use during a crucial driving maneuver. *Accident Analysis and Prevention, 35*, 501–514.

Hancock, P. A., Oron-Gilad, T., and Szalma, J. L. (2007). Elaborations of the multiple-resource theory of attention. In A. F. Kramer, D. A. Wiegmann, and A. Kirlik (Eds.), *Attention: From theory to practice* (pp. 45–56). Oxford: Oxford University Press.

Hancock, P. A., Simmons, L., Hashemi, L., Howarth, H., and Ranney, T. (1999). The effects of in-vehicle distraction on driver response during a crucial driving maneuver. *Transportation Human Factors, 1*, 295–309.

Harbluk, J. L., and Lalande, S. (2005). Performing email tasks while driving: The impact of speech-based tasks on visual detection. *Proceedings of the Third International Driving Symposium on Human Factors in Driver Assessment, Training and Vehicle Design*, 311–317.

Harbluk, J. L., Noy, Y. I., Trbovich, P. L., and Eizenman, M. (2007). An on-road assessment of cognitive distraction: Impacts on drivers' visual behavior and braking performance. *Accident Analysis and Prevention, 39*, 372–379.

Harrar, V., and Harris, L. R. (2005). Simultaneity constancy: Detecting events with touch and vision. *Experimental Brain Research, 166*, 465–473.

Harris, L., Harrar, V., Jaekl, P., and Kopinska, A. (forthcoming). Mechanisms of simultaneity constancy. To appear in R. Nijhawan (Ed.), *Issues of space and time in perception and action.* Cambridge University Press.

Hempel, T., and Altinsoy, E. (2005). Multimodal user interfaces: Designing media for the auditory and the tactile channel. In R. W. Proctor and K.-P. L. Vu (Eds.), *Handbook of human factors in web design* (pp. 134–155). Mahwah, NJ: Lawrence Erlbaum.

Hennessy, D. A., and Wiesenthal, D. L. (1999). Traffic congestion, driver stress, and driver aggression. *Aggressive Behavior, 25*, 409–423.

Hills, B. L. (1980). Vision, visibility, and perception in driving. *Perception, 9*, 183–216.

Hirst, S., and Graham, R. (1997). The format and presentation of collision warnings. In Y. I. Noy (Ed.), *Ergonomics and safety of intelligent driver interfaces* (pp. 203–219). Mahwah, NJ: Lawrence Erlbaum.

Ho, C., Mason, O., and Spence, C. (2007a). An investigation into the temporal dimension of the Mozart effect: Evidence from the attentional blink task. *Acta Psychologica. 125*, 117–128.

Ho, C., Reed, N., and Spence, C. (2006a). Assessing the effectiveness of "intuitive" vibrotactile warning signals in preventing front-to-rear-end collisions in a driving simulator. *Accident Analysis and Prevention, 38*, 989–996. [Journal's web site: http://www.elsevier.com]

Ho, C., Reed, N., and Spence, C. (2007b). Multisensory in-car warning signals for collision avoidance. *Human Factors, 49*, 1107–1114. [Journal's web site: http://www.hfes.org]

Ho, C., Santangelo, V., and Spence, C. (submitted). Multisensory warning signals: When spatial correspondence matters.

Ho, C., and Spence, C. (2005a). Assessing the effectiveness of various auditory cues in capturing a driver's visual attention. *Journal of Experimental Psychology: Applied, 11*, 157–174. [Journal's web site: http://www.apa.org]

Ho, C., and Spence, C. (2005b). Olfactory facilitation of dual-task performance. *Neuroscience Letters, 389*, 35–40.

Ho, C., and Spence C. (2006). Verbal interface design: Do verbal directional cues automatically orient visual spatial attention? *Computers in Human Behavior, 22*, 733–748.

Ho, C., and Spence, C. (2007). Head orientation biases tactile localization. *Brain Research, 1144C*, 136–141.

Ho, C., Spence, C., and Tan, H. Z. (2005a). Warning signals go multisensory. *Proceedings of the 11th International Conference on Human-Computer Interaction, 9*, Paper No. 2284, 1–10.

Ho, C., Tan, H. Z., and Spence, C. (2005b). Using spatial vibrotactile cues to direct visual attention in driving scenes. *Transportation Research Part F: Traffic Psychology and Behaviour, 8*, 397–412. [Journal's web site: http://www.elsevier.com]

Ho, C., Tan, H. Z., and Spence, C. (2006b). The differential effect of vibrotactile and auditory cues on visual spatial attention. *Ergonomics, 49*, 724–738. [Journal's web site: http://www.informaworld.com]

Holmes, N. P., and Spence, C. (2005). Multisensory integration: Space, time, and superadditivity. *Current Biology, 15*, R762–R764.

Holmes, N. P., and Spence, C. (2006). Beyond the body schema: Visual, prosthetic, and technological contributions to bodily perception and awareness. In G. Knoblich, I. M. Thomton, M. Grosjean, and M. Shiffrar (Eds.), *Human body perception from the inside out* (pp. 15–64). Oxford: Oxford University Press.

Hommel, B., Pratt, J., Colzato, L., and Godijn, R. (2001). Symbolic control of visual attention. *Psychological Science, 12*, 360–365.

Hopfinger, J. B., and West, V. M. (2006). Interactions between endogenous and exogenous attention on cortical visual processing. *NeuroImage, 31*, 774–789.

Horberry, T., Anderson, J., Regan, M. A., Triggs, T. J., and Brown, J. (2006). Driver distraction: The effects of concurrent in-vehicle tasks, road environment complexity and age on driving performance. *Accident Analysis and Prevention, 38*, 185–191.

Horne, J. A., and Reyner, L. A. (1995). Sleep related vehicle accidents. *British Journal of Medicine, 310*, 565–567.

Horowitz, A. D., and Dingus, T. A. (1992). Warning signal design: A key human factors issue in an in-vehicle front-to-rear-end collision warning system. *Proceedings of the Human Factors and Ergonomics Society 36th Annual Meeting*, 1011–1013.

Horowitz, M. J., Duff, D. F., and Stratton, L. O. (1964). Body-buffer zone: Exploration of personal space. *Archive of General Psychiatry, 11*, 651–656.

Horrey, W. J., and Wickens, C. D. (2006). Examining the impact of cell phone conversations on driving using meta-analytic techniques. *Human Factors, 48*, 196–205.

Horrey, W. J. (forthcoming). On allocating the eyes: Visual attention and in-vehicle technologies. To appear in C. Castro and L. Hartley (Eds.), *Human factors of visual performance in driving*. Boca Raton, FL: CRC Press.

Horswill, M. S., and McKenna, F. P. (1999). The effect of interference on dynamic risk-taking judgments. *British Journal of Psychology, 90*, 189–199.

Hove, P., Gibbs, G. H., and Caird, J. K. (2000). Identification of positive and negative behaviors associated with using a cellular or mobile phone while driving. In *Proceedings of the Presentation at the Fourteenth Triennial Meeting of the International Ergonomics Association*. San Diego, CA.

Hublet, C., Morais, J., and Bertelson, P. (1976). Spatial constraints on focused attention: Beyond the right-side advantage. *Perception, 5*, 3–8.

Hublet, C., Morais, J., and Bertelson, P. (1977). Spatial effects in speech perception in the absence of spatial competition. *Perception, 6*, 461–466.

Humphrey, C. E. (1952). *Auditory displays: I. Spatial orientation by means of auditory signals – An exploratory study* (Report No. APL/JHU-TG-122). Silver Spring, Maryland: The Johns Hopkins University, Applied Physics Laboratory.

Hunter, W. W., Bundy, H. L., and Daniel, R. B. (1976). *An assessment of the effectiveness of the following-too-closely monitor*. Chapel Hill, NC: University of North Carolina, Highway Safety Research Center.

Hunton, J., and Rose, J. M. (2005). Cellular telephones and driving performance: The effects of attentional demands on motor vehicle crash risk. *Risk Analysis, 25*, 855–866.

Hutton, R., and Smith, E. (2005, 23 October). Self-steering car glides to showrooms. *The Sunday Times*, p. 13.

Hyde, I. H. (1924). Effects of music upon electrocardiograms and blood pressure. *Journal of Experimental Psychology, 7*, 213–224.

Ilmberger, J., Heuberger, E., Mahrhofer, C., Dessovic, H., Kowarik, D., and Buchbauer, G. (2001). The influence of essential oils on human attention. I: Alertness. *Chemical Senses, 26*, 239–245.

Irwin, M., Fitzgerald, C., and Berg, W. P. (2000). Effect of the intensity of wireless telephone conversations on reaction time in a braking response. *Perceptual and Motor Skills, 90*, 1130–1134.

Jackson, M., and Selcon, S. J. (1997). A parallel distributed processing model of redundant information integration. In D. Harris (Ed.), *Engineering psychology and cognitive ergonomics, Vol. 2: Job design and product design* (pp. 193–200). Aldershot, England: Ashgate.

Jagacinski, R. J., Miller, D. P., and Gilson, R. D. (1979). A comparison of kinesthetic-tactual and visual displays via a critical tracking task. *Human Factors, 21*, 79–86.

James, H. F. (1991). Under-reporting of traffic accidents. *Traffic and Engineering Control, 32*, 574–583.

Jamson, A. H., Westerman, S. J., Hockey, G. R. J., and Carstens, O. M. J. (2004). Speech-based e-mail and driver behavior: Effects of an in-vehicle message system interface. *Human Factors, 46*, 625–639.

Jäncke, L., Musial, F., Vogt, J., and Kalveram, K. T. (1994). Monitoring radio programs and time of day affect simulated car-driving performance. *Perceptual and Motor Skills, 79*, 484–486.

Janssen, W., and Nilsson, L. (1993). Behavioural effects of driver support. In A. M. Parkes and S. Franzen (Eds.), *Driving future vehicles* (pp. 147–155). London: Taylor and Francis.

Janssen, W. H., Michon, J. A., and Harvey, L. O., Jr. (1976). The perception of lead vehicle movement in darkness. *Accident Analysis and Prevention, 8*, 151–166.

Janssen, W. H., and Thomas, H. (1997). In-vehicle collision avoidance support under adverse visibility conditions. In Y. I. Noy (Ed.), *Ergonomics and safety of intelligent driver interfaces* (pp. 221–229). Mahwah, NJ: Lawrence Erlbaum.

Jenness, J. W., Lattanzio, R. J., O'Toole, M., Taylor, N., and Pax, C. (2002). Effects of manual versus voice-activated dialing during simulated driving. *Perceptual and Motor Skills, 94*, 363–379.

Johal, S., Napier, F., Britt-Compton, J., and Marshall, T. (2005). Mobile phones and driving. *Journal of Public Health, 27*, 112–113.

Johnson, T. L., and Shapiro, K. L. (1989). Attention to auditory and peripheral visual stimuli: Effects of arousal and predictability. *Acta Psychologica, 72*, 233–245.

Jolicouer, P. (1999). Dual-task interference and visual encoding. *Journal of Experimental Psychology: Human Perception and Performance, 25*, 596–616.

Jones, C. M., Gray, R., Spence, C., and Tan, H. Z. (2008). Directing visual attention with spatially informative and noninformative tactile cues. *Experimental Brain Research, 186*, 659–669.

Jones, L., and Sarter, N. (forthcoming). Tactile displays: Guidance for their design and application. *Human Factors.*

Jordan, P. W., and Johnson, G. I. (1993). Exploring mental workload via TLX: The case of operating a car stereo while driving. In A. G. Gale, I. D. Brown, C. M. Haslegrave, H. W. Kruysse, and S. P. Taylor (Eds.), *Vision in vehicles – IV* (pp. 255–262). Amsterdam: Elsevier Science.

Jordan, T. C. (1972). Characteristics of visual and proprioceptive response times in the learning of a motor skill. *Quarterly Journal of Experimental Psychology, 24*, 536–543.

Just, M. A., Keller, T. A., and Cynkar, J. (forthcoming). A decrease in brain activation associated with driving when listening to someone speak. *Brain Research.*

Kahneman, D., Ben-Ishai, R., and Lotan, M. (1973). Relation of a test of attention to road accidents. *Journal of Applied Psychology, 58,* 113–115.

Kamenetsky, S. B., Hill, D. S., and Trehub, S. E. (1997). Effect of tempo and dynamics on the perception of emotion in music. *Psychology of Music, 25,* 149–160.

Kames, A. J. (1978). A study of the effects of mobile telephone use and control unit design on driving performance. *IEEE Transactions on Vehicular Technology, VT-24,* 282–287.

Kantowitz, B. H., Hanowski, R. H., and Tijerina, L. (1996). Simulator evaluation of heavy-vehicle driver workload II: Complex secondary tasks. *Proceedings of the Human Factors and Ergonomics Society 40th Annual Meeting,* 877–881.

Kantowitz, B. H., Triggs, T. J., and Barnes, V. E. (1990). Stimulus-response compatibility and human factors. In R. W. Proctor and T. G. Reeve (Eds.), *Stimulus-response compatibility: An integrated perspective* (pp. 365–388). Amsterdam: North-Holland.

Karageorghis, C. I., Jones, L., and Low, D. C. (2006). Relationship between exercise heart rate and music tempo preference. *Research Quarterly for Exercise and Sport, 77,* 240–250.

Kawano, T., Iwaki, S., Azuma, Y., Moriwaki, T., and Hamada, T. (2005). Degraded voices through mobile phones and their neural effects: A possible risk of using mobile phones during driving. *Transportation Research Part F, 8,* 331–340.

Kaye, J. J. (2004). Making scents: Aromatic output for HCI. *Interactions, 11,* 48–61.

Kemeny, A., and Panerai, F. (2003). Evaluating perception in driving simulation experiments. *Trends in Cognitive Sciences, 7,* 31–37.

Kennett, S., Eimer, M., Spence, C., and Driver, J. (2001). Tactile-visual links in exogenous spatial attention under different postures: Convergent evidence from psychophysics and ERPs. *Journal of Cognitive Neuroscience, 13,* 462–478.

Kennett, S., Spence, C., and Driver, J. (2002). Visuo-tactile links in covert exogenous spatial attention remap across changes in unseen hand posture. *Perception and Psychophysics, 64,* 1083–1094.

Kenny, T., Anthony, D., Charissis, V., Darawish, Y., and Keir, P. (2004). Integrated vehicle instrument simulations – i-ViS initial design philosophy. *IMechE 2004.*

Kiefer, R., LeBlanc, D., Palmer, M., Salinger, J., Deering, R., and Shulman, M. (1999). *Development and validation of functional definitions and evaluation procedures for collision warning/avoidance systems* (Report No. DOT HS 808 964). Washington, DC: National Highway Transportation Safety Administration.

King, A. J., and Palmer, A. R. (1985). Integration of visual and auditory information in bimodal neurons in the guinea-pig superior colliculus. *Experimental Brain Research, 60,* 492–500.

King, R. A., and Corso, G. M. (1993). Auditory displays: If they are so useful, why are they turned off? *Proceedings of the Human Factors and Ergonomics Society 37th Annual Meeting,* 549–553.

Kinsbourne, M., and Cook, J. (1971). Generalized and lateralized effects of concurrent verbalization on a unimanual skill. *Quarterly Journal of Experimental Psychology, 23,* 341–345.

Kitagawa, N., and Spence, C. (2005). Investigating the effect of a transparent barrier on the crossmodal congruency effect. *Experimental Brain Research, 161*, 62–71.

Kitagawa, N., and Spence, C. (2006). Audiotactile multisensory interactions in human information processing. *Japanese Psychological Research, 48*, 158–173.

Kitagawa, N., Zampini, M., and Spence, C. (2005). Audiotactile interactions in near and far space. *Experimental Brain Research, 166*, 528–537.

Klauer, S. G., Dingus, T. A., Neale, V. L., Sudweeks, J. D., and Ramsey, D. J. (2006). *The impact of driver inattention on near-crash/crash risk: An analysis using the 100-car naturalistic driving study data* (Report No. DOT HS 810 594). Washington, DC: National Highway Transportation Safety Administration.

Klein, R. M. (2004). On the control of visual orienting. In M. I. Posner (Ed.), *Cognitive neuroscience of attention* (pp. 29–44). New York: Guilford Press.

Klein, R. M., and Dick, B. (2002). Temporal dynamics of reflexive attention shifts: A dual-stream rapid serial visual presentation exploration. *Psychological Science, 13*, 176–179.

Klein, R. M., and Shore, D. I. (2000). Relations among modes of visual orienting. In S. Monsell and J. Driver (Eds.), *Attention and performance XVIII: Control of cognitive processes* (pp. 195–208). Cambridge, MA: MIT Press.

Knight, W. (2006, January 12). Clever car keeps an eye on stray pedestrians. *New Scientist*. Retrieved January 17, 2006, from http://www.newscientist.com/article.ns?id=dn8574

Konz, S., and McDougal, D. (1968). The effect of background music on the control activity of an automobile driver. *Human Factors, 10*, 233–244.

Korteling, J. E. (1990). Perception-response speed and driving capabilities of brain-damaged and older drivers. *Human Factors, 32*, 95–108.

Krumhansl, C. L. (2002). Music: A link between cognition and emotion. *Current Directions in Psychological Science, 11*, 45–50.

Kubose, T. T., Bock, K., Dell, G. S., Garnsey, S. M., Kramer, A. F., and Mayhugh, J. (2006). The effects of speech production and speech comprehension on simulated driving performance. *Applied Cognitive Psychology, 20*, 43–63.

Kuc, Z. (1989). A bidirectional vibrotactile communication system: Tactual display design and attainable data rates. *Proceedings of CompEuro '89, VLSI and Computer Peripherals, VLSI and Microelectronic Applications in Intelligent Peripherals and their Interconnection Networks*, 2/101-2/103.

Kume, Y., Shirai, A., Tsuda, M., and Hatada, T. (1998). Information transmission through soles by vibro-tactile stimulation. *Transactions of the Virtual Reality Society of Japan, 3*, 83–88.

Laberge, J., Scialfa, C., White, C., and Caird, J. (2004). The effect of passenger and cellular phone conversations on driver distraction. *Transportation Research Record, 1899*, 109–116.

Laberge-Nadeau, C., Maag, U., Bellavance, F., Lapierre, S. D., Desjardins, D., Messier, S., and Saïdi, A. (2003). Wireless telephones and the risk of road crashes. *Accident Analysis and Prevention, 35*, 649–660.

Làdavas, E. (2002). Functional and dynamic properties of visual peripersonal space. *Trends in Cognitive Sciences, 6*, 17–22.

Làdavas, E., and Farnè, A. (2004). Neuropsychological evidence for multimodal representations of space near specific body parts. In C. Spence and J. Driver (Eds.), *Crossmodal space and crossmodal attention* (pp. 69–98). Oxford: Oxford University Press.

Lamble, D., Kauranen, T., Laasko, M., and Summala, H. (1999). Cognitive load and detection thresholds in car following situations: Safety implications for using mobile (cellular) telephones while driving. *Accident Analysis and Prevention, 31*, 617–623.

Land, M. F. (2006). Eye movements and the control of actions in everyday life. *Progress in Retinal and Eye Research, 25*, 296–324.

Langham, M., Hole, G., Edwards, J., and O'Neil, C. (2002). An analysis of "looked but failed to see" accidents involving parked police cars. *Ergonomics, 45*, 167–185.

Langton, S. R. H., Watt, R. J., and Bruce, V. (2000). Do the eyes have it? Cues to the direction of social attention. *Trends in Cognitive Sciences, 4*, 50–59.

Larsen, L., and Kines, P. (2002). Multidisciplinary in-depth investigations of head-on and left-turn road collisions. *Accident Analysis and Prevention, 34*, 367–380.

Laurienti, P. J., Burdette, J. H., Maldjian, J. A., and Wallace, M. T. (2006). Enhanced multisensory integration in older adults. *Neurobiology of Aging, 27*, 1155–1163.

Lavie, N., and Tsal, Y. (1994). Perceptual load as a major determinant of the locus of selection in visual attention. *Perception and Psychophysics, 56*, 183–197.

Lavie, N. (1995). Perceptual load as a necessary condition for selective attention. *Journal of Experimental Psychology: Human Perception and Performance, 21*, 451–468.

Lavie, N. (2005). Distracted and confused?: Selective attention under load. *Trends in Cognitive Sciences, 9*, 75–82.

Lazarus, H., and Höge, H. (1986). Industrial safety: Acoustic signals for danger situations in factories. *Applied Ergonomics, 17*, 41–46.

Lee, J. D., Caven, B., Haake, S., and Brown, T. L. (2001). Speech-based interaction with in-vehicle computers: The effect of speech-based e-mail on drivers' attention to the roadway. *Human Factors, 43*, 631–640.

Lee, J. D., Gore, B. F., and Campbell, J. L. (1999). Display alternatives for in-vehicle warning and sign information: Message style, location, and modality. *Transportation Human Factors, 1*, 347–375.

Lee, J. D., Hoffman, J. D., and Hayes, E. (2004). Collision warning design to mitigate driver distraction. *Proceedings of the SIGCHI Conference on Human Factors in Computing Systems, 6*, 65–72.

Lee, J. D., McGehee, D. V., Brown, T. L., and Marshall, D. (2006). Effects of adaptive cruise control and alert modality on driver performance. *Transportation Research Record, 1980*, 49–56.

Lee, J. D. (forthcoming). Fifty years of driving safety research. *Human Factors*.

Lee, J. H., and Spence, C. (forthcoming). Assessing the benefits of multimodal feedback on dual-task performance under demanding conditions. *The 22nd British Computer Society Human-Computer Interaction Group Annual Conference*.

Levy, J., Pashler, H., and Boer, E. (2006). Central interference in driving: Is there any stopping the psychological refractory period? *Psychological Science, 17,* 228–235.

Li, W., Moallem, I., Paller, K. A., and Gottfried, J. A. (2007). Subliminal smells can guide social preferences. *Psychological Science, 18,* 1044–1049.

Lindeman, R. W., Yanagida, Y., Sibert, J. L., and Lavine, R. (2003). Effective vibrotactile cueing in a visual search task. *Proceedings of the 9ᵗʰ IFIP TC13 International Conference on Human-Computer Interaction (INTERACT 2003),* 89–96.

Liu, Y.-C. (2003). Effects of Taiwan in-vehicle cellular audio phone system on driving performance. *Safety Science, 41,* 531–542.

Loveland, K. A. (1986). Discovering the affordances of a reflecting surface. *Developmental Review, 6,* 1–24.

Lucas, P. A. (1995). An evaluation of the communicative ability of auditory icons and earcons. *Proceeding of the 2nd International Conference on Auditory Display ICAD '94,* 121–128.

Ludvigson, H. W., and Rottman, T. R. (1989). Effects of ambient odors of lavender and cloves on cognition, memory, affect and mood. *Chemical Senses, 14,* 525–536.

Macaluso, E., Frith, C. D., and Driver, J. (2001). A reply to J. J. McDonald, W. A. Teder-Sälejärvi, and L. M. Ward, Multisensory integration and crossmodal attention effects in the human brain. *Science, 292,* 1791.

Maclure, M., and Mittleman, M. A. (1997). Cautions about car telephones and collisions. *New England Journal of Medicine, 336,* 501–502.

Mariani, M. (2001). COMUNICAR: Designing multimodal interaction for advanced in-vehicle interfaces. *Proceedings of the Human Factors in Transportation, Communication, Health, and the Workplace, Human Factors and Ergonomics Society Europe Chapter Annual Meeting,* 113–120.

Martin, G. N., and Cooper, J. A. (2007). *Odour effects on simulated driving performance: Adding zest to difficult journeys.* Poster presented at the 2007 British Psychology Society Annual Conference, York, England.

Mather, G. (2004). Perceptual and cognitive limits on driver information processing. *IMechE 2004,* 85–92.

Maycock, G. (1996). Sleepiness and driving: The experience of UK car drivers. *Journal of Sleep Research, 5,* 229–237.

Mazzae, E. N., Ranney, T. A., Watson, G. S., and Wightman, J. A. (2004). Hand-held or hands-free? The effects of wireless phone interface type on phone task performance and driver preference. *Proceedings of the 48ᵗʰ Annual Meeting of the Human Factors and Ergonomics Society,* 2218–2222.

McCarley, J. S., Vais, M. J., Pringle, H., Kramer, A. F., Irwin, D. E., and Strayer, D. L. (2004). Conversation disrupts change detection in complex traffic scenes. *Human Factors, 46,* 424–436.

McCartt, A. T., Braver, E. R., and Geary, L. L. (2003). Drivers' use of handheld cell phones before and after New York State's cell phone law. *Preventative Medicine, 36,* 629–635.

McCartt, A. T., and Geary, L. L. (2004). Longer term effects of New York State's law on drivers' handheld cell phone use. *Injury Prevention, 10*, 11–15.

McCartt, A. T., Hellinga, L. A., and Bratiman, K. A. (2006). Cell phones and driving: Review of research. *Traffic Injury Prevention, 7*, 89–106.

McDonald, J. J., Teder-Sälejärvi, W. A., and Ward, L. M. (2001). Multisensory integration and crossmodal attention effects in the human brain. *Science, 292*, 1791.

McEvoy, S. P., Stevenson, M. R., McCartt, A. T., Woodward, M., Haworth, C., Palamara, P., and Cercarelli, R. (2005). Role of mobile phones in motor vehicle crashes resulting in hospital attendance: A case-crossover study. *BMJ 2005*, doi:10.1136/bmj.38537.397512.55

McEvoy, S., Stevenson, M., and Woodward, M. (2007a). The prevalence of, and factors associated with, serious crashes involving a distracting activity. *Accident Analysis and Prevention, 39*, 475–482.

McEvoy, S. P., Stevenson, M. R., and Woodward, M. (2007b). The contribution of passengers versus mobile phone use to motor vehicle crashes resulting in hospital attendance by the driver. *Accident Analysis and Prevention, 39*, 1170–1176.

McGehee, D. V., Brown, T. L., Lee, J. D., and Wilson, T. B. (2002). The effect of warning timing on collision avoidance behavior in a stationary lead vehicle scenario. In *Human performance: Models, intelligent vehicle initiative, traveller advisory and information systems* (pp. 1–7). Washington, DC: Transportation Research Board.

McGehee, D. V., and Raby, M. (2003). *The design and evaluation of snowplow lane awareness system* (Report No. HF-PPC 2003-1). Iowa City: University of Iowa Public Policy Center.

McKeown, J. D. (2005). Candidates for within-vehicle auditory displays. *Proceedings of 11th Meeting of the International Conference on Auditory Display*, 182–189.

McKeown, J. D., and Isherwood, S. (2007). Mapping candidate within-vehicle auditory displays to their referents. *Human Factors, 49*, 417–428.

McKnight, A. J., and McKnight, A. S. (1993). The effect of cellular phone use upon driver attention. *Accident Analysis and Prevention, 25*, 259–265.

McLane, R. C., and Wierwille, W. W. (1975). The influence of motion and audio cues on driver performance in an automobile simulator. *Human Factors, 17*, 488–501.

Menning, H., Ackermann, H., Hertrich, I., and Mathiak, K. (2005). Spatial auditory attention is modulated by tactile priming. *Experimental Brain Research, 164*, 41–47.

Michael, G. A., Jacquot, L., Millot, J.-L., and Brand, G. (2003). Ambient odors modulate visual attentional capture. *Neuroscience Letters, 352*, 221–225.

Michelitsch, G., Williams, J., Osen, M., Jimenez, B., and Rapp, S. (2004). Haptic chameleon: A new concept of shape-changing user interface controls with force feedback. *CHI '04 Extended Abstracts on Human Factors in Computing Systems*, 1305–1308.

Mihal, W. L., and Barrett, G. V. (1976). Individual differences in perceptual information processing and their relation to automobile accident involvement. *Journal of Applied Psychology, 61*, 229–233.

Mikkonen, V., and Backman, M. (1988). *Use of the car telephone while driving.* Technical report no. A 39. Helsinki: Department of Psychology, University of Helsinki.

Miller, E. (1970). Simple and choice reaction time following severe head injury. *Cortex, 6,* 121–127.

Miller, J. (1982). Divided attention: Evidence for coactivation with redundant signals. *Cognitive Psychology, 14,* 247–279.

Millot, J.-L., Brand, G., and Morand, N. (2002). Effects of ambient odors on reaction time in humans. *Neuroscience Letters, 322,* 79–82.

Miltner, W., Matjak, M., Braun, C., Diekmann, H., and Brody, S. (1994). Emotional qualities of odors and their influence on the startle reflex in humans. *Psychophysiology, 31,* 107–110.

Montagu, A. (1971). *Touching: The human significance of the skin.* New York: Columbia University Press.

Moore, D. R., and King, A. J. (1999). Auditory perception: The near and far of sound localization. *Current Biology, 9,* R361–R363.

Morais, J. (1978). Spatial constraints on attention to speech. In J. Requin (Ed.), *Attention and performance VII* (pp. 245–260). Hillsdale, NJ: LEA.

Mordkoff, J. T., and Egeth, H. E. (1993). Response time and accuracy revisited: Converging support for the interactive race model. *Journal of Experimental Psychology: Human Perception and Performance, 19,* 981–991.

Morris, R. W., and Montano, S. R. (1996). Response times to visual and auditory alarms during anaesthesia. *Anaesthesia and Intensive Care, 24,* 682–684.

Moss, M., Cook, J., Wesnes, K., and Duckett, P. (2003). Aromas of rosemary and lavender essential oils differentially affect cognition and mood in healthy adults. *International Journal of Neuroscience, 113,* 15–38.

Motter, B. (2001). Attention in the animal brain. In R. A. Wilson and F. C. Keil (Eds.), *The MIT encyclopedia of the cognitive sciences* (pp. 41–43). Cambridge, MA: MIT Press.

Mouloua, M., Hancock, P. A., Rinalducci, E., and Brill, C. (2003). Effects of radio tuning on driving performance. *Proceedings of the Human Factors and Ergonomics Society 47th Annual Meeting,* 1044–1047.

Mowbray, G. H., and Gebhard, J. W. (1961). Man's senses as informational channels. In H. W. Sinaiko (Ed.), *Selected papers on human factors in the design and use of control systems* (pp. 115–149). New York: Dover.

Murray, J., Ayres, T., Wood, C., and Humphrey, D. (2001). Mobile communications, driver distraction and vehicle accidents. *International Journal of Vehicle Design, 26,* 70–84.

Murray, M. M., Molholm, S., Michel, C. M., Heslenfeld, D. J., Ritter, W., Javitt, D. C., et al. (2005). Grabbing your ear: Rapid auditory-somatosensory multisensory interactions in low-level sensory cortices are not constrained by stimulus alignment. *Cerebral Cortex, 15,* 963–974.

Nagel, S. K., Carl, C., Kringe, T., Martin, R., and Konig, P. (2005). Beyond sensory substitution – Learning the sixth sense. *Journal of Neural Engineering, 2,* R13–R26.

Navarra, J., Alsius, A., Soto-Faraco, S., and Spence, C. (forthcoming). Is attention involved in the audiovisual integration of speech? *Information Fusion*.

Navon, D. (1984). Resources: A theoretical soup stone? *Psychological Review, 91*, 216–234.

Nelson, R. J., McCandlish, C. A., and Douglas, V. D. (1990). Reaction times for hand movements made in response to visual versus vibratory cues. *Somatosensory and Motor Research, 7*, 337–352.

Nilsson, L. (1993). Contributions and limitations of simulator studies to driver behaviour research. In A. M. Parkes and S. Franzen (Eds.), *Driving future vehicles* (pp. 401–407). London: Taylor and Francis.

North, A. C., and Hargreaves, D. J. (1999). Music and driving game performance. *Scandinavian Journal of Psychology, 40*, 285–292.

North, A. C., Hargreaves, D. J., and Hargreaves, J. J. (2004). Uses of music in everyday life. *Music Perception, 22*, 41–77.

Noy, Y. I. (1997). Human factors in modern traffic systems. *Ergonomics, 40*, 1016–1024.

Nunes, L. M., and Recarte, M. A. (2002). Cognitive demands of hands-free-phone conversation while driving. *Transportation Research, Part F, 5*, 133–144.

O'Connell, R. J., Stevens, D. A., and Zogby, L. M. (1994). Individual differences in the perceived intensity and quality of specific odors following self- and cross-adaptation. *Chemical Senses, 19*, 197–208.

Oh, E. L., and Lutfi, R. A. (1999). Informational masking by everyday sounds. *Journal of the Acoustical Society of America, 106*, 3521–3528.

Oldham, G. R., Cummings, A., Mischel, L. J., Schmidtke, J. M., and Zhou, J. (1995). Listen while you work? Quasi-experimental relations between personal-stereo headset use and employee work responses. *Journal of Applied Psychology, 80*, 547–564.

O'Regan, J. K., Rensink, R. A., and Clark, J. J. (1999). Change-blindness as a result of 'mudsplashes'. *Nature, 398*, 34.

Oron-Gilad, T., Ronen, A., and Shinar, D. (2008). Alertness maintaining tasks (AMTs) while driving. *Accident Analysis and Prevention, 40*, 851–860.

Oron-Gilad, T., and Shinar, D. (2000). Driver fatigue among military truck drivers. *Transportation Research Part F, 3*, 195–209.

Otmani, S., Pebayle, T., Roge, J., and Muzet, A. (2005). Effect of driving duration and partial sleep deprivation on subsequent alertness and performance of car drivers. *Physiology and Behavior, 84*, 715–724.

Otten, L. J., Alain, C., and Picton, T. W. (2000). Effects of visual attentional load on auditory processing. *NeuoReport, 11*, 875–880.

Oviatt, S. (1999). Ten myths of multimodal interaction. *Communications of the ACM, 42*, 74–81.

Oviatt, S. (2002). Multimodal interfaces. In J. A. Jacko and A. Sears (Eds.), *The human-computer interaction handbook: Fundamentals, evolving technologies and emerging applications* (pp. 286–304). Mahwah, NJ: Lawrence Erlbaum.

Oyer, J., and Hardick, J. (1963). *Response of population to optimum warning signal* (Report No. SHSLR163). Washington, DC: Office of Civil Defence.

Paillard, J. (1993). The hand and the tool: The functional architecture of human technical skills. In A. Berthelet and J. Chavaillon (Eds.), *The use of tools by human and non-human primates* (pp. 34–46). New York: Oxford University Press.

Panksepp, J. (1995). The emotional sources of chills induced by music. *Music Perception, 13*, 171–207.

Parasuraman, R., Hancock, P. A., and Olofinboba, O. (1997). Alarm effectiveness in driver-centred collision-warning systems. *Ergonomics, 40*, 390–399.

Parasuraman, R., and Riley, V. (1997). Humans and automation: Use, misuse, disuse, abuse. *Human Factors, 39*, 230–253.

Parkes, A. M. (1991a). Drivers' business decision making ability whilst using carphones. In E. J. Lovesey (Ed.), *Contemporary Ergonomics 1991: Proceedings of the Ergonomics Society's 1991 Annual Conference* (pp. 427–432). London: Taylor and Francis.

Parkes, A. M. (1991b). The effects of driving and handsfree telephone use on conversation structure and style. *Proceedings of Human Factors Association of Canada Conference*, 141–147.

Pashler, H. (1991). Shifting visual attention and selecting motor responses: Distinct attentional mechanisms. *Journal of Experimental Psychology: Human Perception and Performance, 17*, 1023–1040.

Patten, C. J. D., Kircher, A., Ostlund, J., and Nilsson, L. (2004). Using mobile telephones: Cognitive workload and attention resource allocation. *Accident Analysis and Prevention, 36*, 341–350.

Peltzman, S. (1975). The effects of automobile safety regulation. *Journal of Political Economy, 83*, 677–725.

Perrott, D. R., Cisneros, J., McKinley, R. L., and D'Angelo, W. R. (1996). Aurally aided visual search under virtual and free-field listening conditions. *Human Factors, 38*, 702–715.

Perrott, D. R., Saberi, K., Brown, K., and Strybel, T. Z. (1990). Auditory psychomotor coordination and visual search performance. *Perception and Psychophysics, 48*, 214–226.

Perrott, D. R., Sadralodabai, T., Saberi, K., and Strybel, T. Z. (1991). Aurally aided visual search in the central visual field: Effects of visual load and visual enhancement of the target. *Human Factors, 33*, 389–400.

Peters, G. A., and Peters, B. J. (2001). The distracted driver. *Journal of the Royal Society for the Promotion of Health, 121*, 23–28.

Peterson, S., Hoffer, G., and Millner, E. (1995). Are drivers of air-bag-equipped cars more aggressive? A test of the offsetting behavior hypothesis. *Journal of Law and Economics, 38*, 251–264.

Petica, S. (1993). Risks of cellular phone usage in the car and its impact on road safety. *Recherche-Transporte-Securite, 37*, 45–56.

'Phones and automobiles can be a dangerous mix' (1999). *State Legislatures, 25(4)*, 5.

Piersma, E. H. (1993). Adaptive interfaces and support systems in future vehicles. In A. M. Parkes and S. Franzen (Eds.), *Driving future vehicles* (pp. 321–332). London: Taylor and Francis.

Ponds, R. W. H. M., Brouwer, W. H., and van Wolffelaar, P. C. (1988). Age differences in divided attention in a simulated driving task. *Journal of Gerontology: Psychological Sciences, 43*, P151–P156.

Pöppel, E. (1988). *Mindworks: Time and conscious experience.* Boston: Harcourt.

Posner, M. I. (1978). *Chronometric explorations of mind.* Hillsdale, NJ: Erlbaum.

Posner, M. I. (1980). Orienting of attention. *Quarterly Journal of Experimental Psychology, 32*, 3–25.

Posner, M. I., and Boies, S. J. (1971). Components of attention. *Psychological Review, 78*, 391–408.

Posner, M. I., Nissen, M. J., and Klein, R. M. (1976). Visual dominance: An information-processing account of its origins and significance. *Psychological Review, 83*, 157–171.

Previc, F. H. (1998). The neuropsychology of 3-D space. *Psychological Bulletin, 124*, 123–164.

Previc, F. H. (2000). Neuropsychological guidelines for aircraft control stations. *IEEE Engineering in Medicine and Biology Magazine, 19*, 81–88.

Prinzmetal, W., McCool, C., and Park, S. (2005). Attention: Reaction time and accuracy reveal different mechanisms. *Journal of Experimental Psychology: General, 134*, 73–92.

Proctor, R. W., Tan, H. Z., Vu, K.-P. L., Gray, R., and Spence, C. (2005). Implications of compatibility and cuing effects for multimodal interfaces. *Proceedings of the 11th International Conference on Human-Computer Interaction, 11*, Paper No. 2733, 1–10.

Quinlan, K. P. (1997). Commentary on cellular telephones and traffic accidents. *New England Journal of Medicine, 337*, 127–129.

RAC Foundation. (2005, June 3). *The scent of danger.* Retrieved June 4, 2005, from http://www.racfoundation.org/releases/030605rac.htm

Rakauskas, M., Gugerty, L., and Ward, N. J. (2004). Effects of cell phone conversations on driving performance. *Journal of Safety Research, 35*, 453–464.

Ramsey, K. L., and Simmons, F. B. (1993). High-powered automobile stereos. *Otolaryngology Head and Neck Surgery, 109*, 108–110.

Raudenbush, B., Corley, N., and Eppich, W. (2001). Enhancing athletic performance through the administration of peppermint odor. *Journal of Sport and Exercise Psychology, 23*, 156–160.

Rauscher, F. H., Shaw, G. L., and Ky, K. N. (1993). Music and spatial task performance. *Nature, 365*, 611.

Recarte, M. A., and Nunes, L. M. (2000). Effects of verbal and spatial-imagery tasks on eye fixations while driving. *Journal of Experimental Psychology: Applied, 6*, 31–43.

Recarte, M. A., and Nunes, L. M. (2003). Mental workload while driving: Effects on visual search, discrimination, and decision making. *Journal of Experimental Psychology: Applied, 9*, 119–137.

Redelmeier, D. A., and Tibshirani, R. J. (1997a). Association between cellular-telephone calls and motor vehicle collisions. *New England Journal of Medicine, 336*, 453–458.

Redelmeier, D. A., and Tibshirani, R. J. (1997b). Commentary on cellular telephones and traffic accidents. *New England Journal of Medicine, 337,* 127–129.

Redelmeier, D. A., and Weinstein, M. C. (1999). Cost-effectiveness of regulations against using a cellular telephone while driving. *Medical Decision Making, 19,* 1–8.

Reed, M. P., and Green, P. A. (1999). Comparison of driving performance on-road and in a low-cost simulator using a concurrent telephone dialling task. *Ergonomics, 42,* 1015–1037.

Rees, G., and Lavie, N. (2001). What can functional imaging reveal about the role of attention in visual awareness? *Neuropsychologia, 39,* 1343–1353.

Rensink, R. A. (2004). Visual sensing without seeing. *Psychological Science, 15,* 27–32.

Rensink, R. A., O'Regan, J. K., and Clark, J. J. (1997). To see or not to see: The need for attention to perceive changes in scenes. *Psychological Science, 8,* 368–373.

Rentfrow, P. J., and Gosling, S. D. (2003). The do re mi's of everyday life: The structure and personality correlates of music preferences. *Journal of Personality and Scoial Psychology, 84,* 1236–1256.

Reyner, L. A., and Horne, J. A. (1998). Falling asleep whilst driving: Are drivers aware of prior sleepiness? *International Journal of Legal Medicine, 111,* 120–123.

Richard, C. M., Wright, R. D., Ee, C., Prime, S. L., Shimizu, Y., and Vavrik, J. (2002). Effect of a concurrent auditory task on visual search performance in a driving-related image-flicker task. *Human Factors, 44,* 108–119.

Riddoch, G. (1941). Phantom limbs and body shape. *Brain, 64,* 197–222.

Rinspeed "Senso" – The car that senses the driver. (n.d.). Retrieved May 30, 2005, from http://www.rinspeed.com/pages/cars/senso/pre-senso.htm

Rizzolatti, G., Fadiga, L., Fogassi, L., and Gallese, V. (1997). The space around us. *Science, 277,* 190–191.

Rochlis, J. L., and Newman, D. J. (2000). A tactile display for International Space Station (ISS) extravehicular activity (EVA). *Aviation, Space, and Environmental Medicine, 71,* 571–578.

Rockwell, T. H. (1988). Spare visual capacity in driving – revisited: New empirical results for an old idea. In A. G. Gale, M. H. Freeman, C. M. Haslegrave, P. Smith, and S. P. Taylor (Eds.), *Vision in vehicles II* (pp. 317–324). Amsterdam: Elsevier Science.

Rodway, P. (2005). The modality shift effect and the effectiveness of warning signals in different modalities. *Acta Psychologica, 120,* 199–226.

Rogers, R. D., and Monsell, S. (1995). Costs of a predictive switch between simple cognitive tasks. *Journal of Experimental Psychology: General, 124,* 207–231.

Rosenbloom, T. (2006). Driving performance while using cell phones: An observational study. *Journal of Safety Research, 37,* 207–212.

Royal, D. (2003). *Volume 1: Findings. National survey of distracted and drowsy driving attitudes and behaviors, 2002.* National Highway Traffic Safety Administration (NHTSA) Report Number: DOT HS 809 566.

Rumar, K. (1990). The basic driver error: Late detection. *Ergonomics, 33*, 1281–1290.

Rupert, A. H. (2000a). An instrumentation solution for reducing spatial disorientation mishaps. *IEEE Engineering in Medicine and Biology Magazine, 19*, 71–80.

Rupert, A. H. (2000b). Tactile situation awareness system: Proprioceptive prostheses for sensory deficiencies. *Aviation, Space, and Environmental Medicine, 71*, A92–A99.

Rupert, A. H., Guedry, F. E., and Reschke, M. F. (1994). *The use of a tactile interface to convey position and motion perceptions* (Report No. AGARD-CP-541-20(1)-20-(7)). France: Advisory Group for Aerospace Research and Development.

Sagberg, F. (1999). Road accidents caused by drivers falling asleep. *Accident Analysis and Prevention, 31*, 639–649.

Sagberg, F. (2001). Accident risk of car drivers during mobile telephone use. *International Journal of Vehicle Design, 26*, 57–69.

Salvucci, D. D. (2001). Predicting the effects of in-car interface use on driver performance: An integrated model approach. *International Journal of Human-Computer Studies, 55*, 85–107.

Salvucci, D. D., and Macuga, K. L. (2002). Predicting the effects of cellular-phone dialing on driver performance. *Cognitive Systems Research, 3*, 95–102.

Sample, I. (2001). You drive me crazy. *New Scientist, 171 (2300)*, 24.

Santangelo, V., Belardinelli, M. O., and Spence, C. (2007). The suppression of reflexive visual and auditory orienting when attention is otherwise engaged. *Journal of Experimental Psychology: Human Perception and Performance, 33*, 137–148.

Santangelo, V., and Spence, C. (2007). Multisensory cues capture spatial attention regardless of perceptual load. *Journal of Experimental Psychology: Human Perception and Performance, 33*, 1311–1321.

Santangelo, V., and Spence, C. (forthcoming). Is the exogenous orienting of spatial attention truly automatic? A multisensory perspective. *Consciousness and Cognition*.

Sarter, N. B. (2006). Multimodal information presentation: Design guidance and research challenges. *International Journal of Industrial Ergonomics, 36*, 439–445.

Sarter, N., and Sarter, M. (2003). Neuroergonomics: Opportunities and challenges of merging cognitive neuroscience with cognitive ergonomics. *Theoretical Issues in Ergonomics Science, 4*, 142–150.

Schellenberg, E. G. (2001). Music and nonmusical abilities. *Annals of the New York Academy of Sciences, 930*, 355–371.

Schumacher, E. H., Seymour, T. L., Glass, J. M., Fencsik, D. E., Lauber, E. J., Kieras, D. E., et al. (2001). Virtually perfect time sharing in dual-task performance: Uncorking the central cognitive bottleneck. *Psychological Science, 12*, 101–108.

Schumacher, R., et al. (2007). *Macht Mozart schlau? Die Förderung kognitiver Kompetenzen durch Musik.* [Does Mozart make smart? The promotion of cognitive authority by music.] Retrieved 14 April 2007, from http://www.bmbf.de/pub/macht_mozart_schlau.pdf

Schumann, J., Godthelp, H., Farber, B., and Wontorra, H. (1993). Breaking up open-loop steering control actions the steering wheel as an active control device. In A. G. Gale, I. D. Brown, C. M. Haslegrave, H. W. Kruysse, and S. P. Taylor (Eds.), *Vision in vehicles – IV* (pp. 321–332). Amsterdam: Elsevier Science.

Schumann, J., Godthelp, J., and Hoekstra, W. (1992). *An exploratory simulator study on the use of active control devices in car driving* (Report No. IZF 1992 B-2). Soesterberg, Netherlands: TNO Institute for Perception.

Schumann, J., and Naab, K. (1992). On the effectiveness of an active steering wheel in critical driving situations – A proving ground experiment. *Proceedings of the Conference on Road Safety in Europe*, 194–208.

Selcon, S. J., Taylor, R. M., and McKenna, F. P. (1995). Integrating multiple information sources: Using redundancy in the design of warnings. *Ergonomics, 38*, 2362–2370.

Selcon, S. J., Taylor, R. M., and Shadrake, R. A. (1991). Giving the pilot two sources of information: Help or hindrance. In E. Farmer (Ed.), *Human resource management in aviation* (pp. 139–148). Aldershot, England: Avebury Technical.

Senders, J. W., Kristofferson, A. B., Levison, W. H., Dietrich, C. W., and Ward, J. L. (1967). The attentional demand of automobile driving. *Highway Research Record, 195*, 15–33.

Serafin, C., Wen, C., Paelke, G., and Green, P. (1993a). *Development and human factors tests of car phones.* Technical Report No. UMTRI-93-17. The University of Michigan Transportation Research Institute, Ann Arbor, MI.

Serafin, C., Wen, C., Paelke, G., and Green, P. (1993b). Car phone usability: A human factors laboratory test. *Proceedings of the Human Factors and Ergonomics Society 37th Annual Meeting*, 220–224.

Shapiro, K. (Ed.). (2001). *The limits of attention: Temporal constraints in human information processing.* Oxford: Oxford University Press.

Shinar, D. (1978). *Psychology on the road: The human factor in traffic safety.* New York: Wiley.

Shinar, D., Tractinsky, N., and Compton, R. (2005). Effects of practice, age, and task demands, on interference from a phone task while driving. *Accident Analysis and Prevention, 37,* 315–326.

Simons, D. J., and Chabris, C. F. (1999). Gorillas in our midst: Sustained inattentional blindness for dynamic events. *Perception, 28*, 1059–1074.

Simons, D. J., and Rensink, R. A. (2005). Change blindness: Past, present, and future. *Trends in Cognitive Sciences, 9*, 16–20.

Simpson, C. A., and Marchionda-Frost, K. (1984). Synthesized speech rate and pitch effects on intelligibility of warning messages for pilots. *Human Factors, 26*, 509–517.

Simpson, C. A., McCauley, M. E., Roland, E. F., Ruth, J. C., and Williges, B. H. (1987). Speech controls and displays. In G. Salvendy (Ed.), *Handbook of human factors* (pp. 1490–1525). New York: Wiley.

Sinnett, S., Spence, C., and Soto-Faraco, S. (2007). Visual dominance and attention: The Colavita effect revisited. *Perception and Psychophysics, 69,* 673–686.

Sinnett, S., Spence, C., and Soto-Faraco, S. (2008). Theco-occurrence of multisensory competition and facilitation. *Acta Psychologica, 128,* 153–161

Sivak, M. (1996). The information that drivers use: Is it indeed 90% visual? *Perception, 25*, 1081–1090.

Sklar, A. E., and Sarter, N. B. (1999). Good vibrations: Tactile feedback in support of attention allocation and human-automation coordination in event-driven domains. *Human Factors, 41*, 543–552.

Slawinski, E. B., and MacNeil, J. F. (2002). Age, music, and driving performance: Detection of external warning sounds in vehicles. *Psychomusicology, 18*, 123–131.

Sloboda, J. A. (1999). Everyday uses of music listening: A preliminary study. In S. W. Yi (Ed.), *Music, mind, and science* (pp. 354–369). Seoul: Western Music Institute.

Sloboda, J. A., O'Neill, S. A., and Vivaldi, A. (2001). Functions of music in everyday life: An exploratory study using experience sampling method. *Musicae Scientiae, 5*, 9–32.

Smith, D. L., Najm, W. G., and Lam, A. H. (2003). Analysis of braking and steering performance in car-following scenarios. *2003 SAE World Congress*, SAE Technical Paper Series No. 2003-01-0283.

Smith, E. (2008; May 11). 'Don't panic, calamity, the crash-proof car is coming. *The Sunday Times* (InGear), May 11, 4–5.

Smith, R. K., Luke, T., Parkes, A. P., Burns, P. C., and Landsdown, T. C. (2005). A study of driver visual behavior while talking with passengers, and on mobile phones. In D. de Waard, K. A. Brookhuis, R. van Egmond, and Th. Boersema (Eds.), *Human factors in design, safety, and management* (pp. 11–20). Maastricht, The Netherlands: Shaker Publishing

Smith, W. A. (1961). Effects of industrial music in a work situation requiring complex mental activity. *Psychological Reports, 8*, 159–162.

Snyder, J. B., Nelson, M. E., Burdick, J. W., and MacIver, M. A. (2007). Omnidirectional sensory and motor volumes in electric fish. *PLoS Biology, 5*, 2671–2683.

Sommer, R. (1959). Studies in personal space. *Sociometry, 22*, 247–260.

Sorkin, R. D. (1987). Design of auditory and tactile displays. In G. Salvendy (Ed.), *Handbook of human factors* (pp. 549–576). New York: Wiley.

Sorkin, R. D. (1988). Why are people turning off our alarms? *Journal of the Acoustical Society of America, 84*, 1107–1108.

Soto-Faraco, S., Morein-Zamir, S., and Kingstone, A. (2005). On audiovisual spatial synergy: The fragility of an effect. *Perception and Psychophysics, 67*, 444–457.

Spence, C. (2001). Crossmodal attentional capture: A controversy resolved? In C. L. Folk and B. S. Gibson (Eds.), *Attraction, distraction and action: Multiple perspectives on attentional capture* (pp. 231–262). Amsterdam: Elsevier Science.

Spence, C. (2002). *The ICI report on the secret of the senses*. London: The Communication Group.

Spence, C. J., and Driver, J. (1994). Covert spatial orienting in audition: Exogenous and endogenous mechanisms. *Journal of Experimental Psychology: Human Perception and Performance, 20*, 555–574.

Spence, C., and Driver, J. (1996). Audiovisual links in endogenous covert spatial attention. *Journal of Experimental Psychology: Human Perception and Performance, 22*, 1005–1030.

Spence, C., and Driver, J. (1997a). Cross-modal links in attention between audition, vision, and touch: Implications for interface design. *International Journal of Cognitive Ergonomics, 1*, 351–373.

Spence, C., and Driver, J. (1997b). On measuring selective attention to a specific sensory modality. *Perception and Psychophysics, 59*, 389–403.

Spence, C., and Driver, J. (1999). A new approach to the design of multimodal warning signals. In D. Harris (Ed.), *Engineering psychology and cognitive ergonomics, Vol. 4: Job design, product design and human-computer interaction* (pp. 455–461). Aldershot, England: Ashgate.

Spence, C., and Driver, J. (2000). Attracting attention to the illusory location of a sound: Reflexive crossmodal orienting and ventriloquism. *NeuroReport, 11*, 2057–2061.

Spence, C., and Driver, J. (Eds.). (2004). *Crossmodal space and crossmodal attention.* Oxford: Oxford University Press.

Spence, C, and Gallace, A. (2007). Recent developments in the study of tactile attention. *Canadian Journal of Experimental Psychology, 61*, 196–207.

Spence, C., and Ho, C. (2008). Multisensory interface design for drivers: Past, present and future. *Ergonomics, 51*, 65–70. [Journal's web site: http://www. informaworld.com]

Spence, C., and Ho, C. (forthcoming). Multisensory warning signals for event perception and safe driving. *Theoretical Issues in Ergonomic Science.* [Journal's web site: http://www.tandf.co.uk]

Spence, C., and Ho, C. (submitted). Tactile and multisensory warning signals for drivers. *IEEE Transactions on Haptics.*

Spence, C., and McDonald, J. (2004). The cross-modal consequences of the exogenous spatial orienting of attention. In G. A. Calvert, C. Spence, and B. E. Stein (Eds.), *The handbook of multisensory processes* (pp. 3–25). Cambridge, MA: MIT Press.

Spence, C., McDonald, J., and Driver, J. (2004). Exogenous spatial cuing studies of human crossmodal attention and multisensory integration. In C. Spence and J. Driver (Eds.), *Crossmodal space and crossmodal attention* (pp. 277–320). Oxford: Oxford University Press.

Spence, C., Nicholls, M. E. R., and Driver, J. (2001). The cost of expecting events in the wrong sensory modality. *Perception and Psychophysics, 63*, 330–336.

Spence, C., Nicholls, M. E. R., Gillespie, N., and Driver, J. (1998). Cross-modal links in exogenous covert spatial orienting between touch, audition, and vision. *Perception and Psychophysics, 60*, 544–557.

Spence, C., Pavani, F., and Driver, J. (2000a). Crossmodal links between vision and touch in covert endogenous spatial attention. *Journal of Experimental Psychology: Human Perception and Performance, 26*, 1298–1319.

Spence, C., Ranson, J., and Driver, J. (2000b). Cross-modal selective attention: On the difficulty of ignoring sounds at the locus of visual attention. *Perception and Psychophysics, 62*, 410–424.

Spence, C., and Read, L. (2003). Speech shadowing while driving: On the difficulty of splitting attention between eye and ear. *Psychological Science, 14*, 251–256.

Spence, C., and Squire, S. (2003). Multisensory integration: Maintaining the perception of synchrony. *Current Biology, 13*, R519–R521.

Stanton, N., and Edworthy, J. (1999). *Human factors in auditory warnings.* Aldershot, England: Ashgate.

Stanton, N. A., and Young, M. S. (2005). Driver behaviour with adaptive cruise control. *Ergonomics, 48*, 1294–1313.

Steele, M., and Gillespie, B. (2001). Shared control between human and machine: Using a haptic steering wheel to aid in land vehicle guidance. *Proceedings of the Human Factors and Ergonomics Society 45ᵗʰ Annual Meeting*, 1671–1675.

Stein, A. C., Parseghian, Z., and Allen, R. W. (1987). A simulator study of the safety implications of cellular mobile phone use. *Proceedings of the 31st Annual Conference of the American Association for Automative Medicine*, 181–200.

Stein, B. E., and Meredith, M. A. (1993). *The merging of the senses*. Cambridge, MA: MIT Press.

Stein, B. E., and Stanford, T. R. (2008). Multisensory integration: Current issues from the perspective of the single neuron. *Nature Reviews Neuroscience, 9*, 255–267.

Stevens, S. S., and Newman, E. B. (1936). The localization of actual sources of sound. *American Journal of Psychology, 48*, 297–306.

Stokes, A., Wickens, C., and Kite, K. (1990). *Display technology: Human factors concepts*. Warrendale, PA: Society of Automotive Engineers.

Strayer, D. L., Cooper, J. M., and Drews, F. A. (2004). What do drivers fail to see when conversing on a cell phone? *Proceedings of the Human Factors and Ergonomics Society 48ᵗʰ Annual Meeting 2004*, 2213–2217.

Strayer, D. L., and Drews, F. A. (2004). Profiles in driver distraction: Effects of cell phone conversations on younger and older drivers. *Human Factors, 46*, 640–649.

Strayer, D. L., and Drews, F. A. (2007). Multitasking in the automobile. In A. F. Kramer, D. A. Wiegmann, and A. Kirlik (Eds.), *Attention: From theory to practice* (pp. 121–133). Oxford University Press.

Strayer, D. L., Drews, F. A., and Crouch, D. J. (2006). A comparison of the cell phone driver and the drunk driver. *Human Factors, 48*, 381–391.

Strayer, D. L., Drews, F. A., and Johnston, W. A. (2003). Cell phone-induced failures of visual attention during simulated driving. *Journal of Experimental Psychology: Applied, 9*, 23–32.

Strayer, D. L., and Johnston, W. A. (2001). Driven to distraction: Dual-task studies of simulated driving and conversing on a cellular telephone. *Psychological Science, 12*, 462–466.

Streeter, L. A., Vitello, D., and Wonsiewicz, S. A. (1985). How to tell people where to go: Comparing navigational aids. *International Journal of Man-Machine Studies, 22*, 549–562.

Stuss, D. T., Stetham, L. L., Hugenholtz, H., Picton, T., Pivik, J., and Richard, M. T. (1989). Reaction time after head injury: Fatigue, divided and focused attention, and consistency of performance. *Journal of Neurology, Neurosurgery, and Psychiatry, 52*, 742–748.

Stutts, J., Feaganes, J., Reinfurt, D., Rodgman, E., Hamlett, C., Gish, K., et al. (2005). Driver's exposure to distractions in their natural driving environment. *Accident Analysis and Prevention, 37*, 1093–1101.

Stutts, J. C., Reinfurt, D. W., and Rodgman, E. A. (2001). The role of driver distraction in crashes: An analysis of 1995–1999 Crashworthiness Data System data. *45th Annual Proceedings Association for the Advancement of Automative Medicine* (pp. 287–301). Des Plaines, IA: AAAM.

Suetomi, T., and Kido, K. (1997). Driver behavior under a collision warning system – A driving simulator study. *SAE Special Publications, 1242*, 75–81.

Sullman, M. J. M., and Baas, P. H. (2004). Mobile phone use amongst New Zealand drivers. *Transportation Research: Part F, 7*, 95–105.

Summala, H. (1997). Ergonomics of road transport. *IATSS Research, 21*, 49–57.

Sundeen, M. (2006). Cell phones and highway safety: 2001 State Legislative update. *National Conference of State Legislatures: Environment, Energy, and Transportation Program*. Retrieved 5th March, 2008, from http://www.ncsl.org/print/transportation/2006cellphone.pdf

Sussman, E. D., Bishop, H., Madnick, B., Walter R. (1985). Driver inattention and highway safety. *Transportation Research Record, 1047*, 40–48.

Suzuki, K., and Jansson, H. (2003). An analysis of driver's steering behaviour during auditory or haptic warnings for the designing of lane departure warning system. *JSAE Review, 24*, 65–70.

Symons, J. R. (1963). The effect of various heteromodal stimuli on visual sensitivity. *Quarterly Journal of Experimental Psychology, 15*, 243–251.

Tajadura-Jiménez, A., Kitagawa, N., Väljamäe, A., Zampini, M., Murray, M. M., and Spence, C. (submitted). Audiotactile interactions in near and far space. *Brain Research*.

Tan, H. Z., Gray, R., Young, J. J., and Traylor, R. (2003). A haptic back display for attentional and directional cueing. *Haptics-e, 3(1)*. Retrieved 11 June 2003, from http://www.haptics-e.org/Vol_03/he-v3n1.pdf

Tannen, R. S., Nelson, W. T., Bolia, R. S., Warm, J. S., and Dember, W. N. (2004). Evaluating adaptive multisensory displays for target localization in a flight task. *International Journal of Aviation Psychology, 14*, 297–312.

Tassinari, G., and Campara, D. (1996). Consequences of covert orienting to non-informative stimuli of different modalities: A unitary mechanism? *Neuropsychologia, 34*, 235–245.

Tellinghuisen, D. J., and Nowak, E. J. (2003). The inability to ignore auditory distractors as a function of visual task perceptual load. *Perception and Psychophysics, 65*, 817–828.

Terry, H. R., Charlton, S. G., and Perrone, J. A. (forthcoming). The role of looming and attention capture in drivers' braking responses. *Accident Analysis and Prevention*.

'The mobile phone report: A report on the effects of using a 'hand-held' and 'hands-free' mobile phone on road safety' (2002). Direct Line Motor Insurance. Retrieved January 27, 2008, fromhttp://info.directline.com/.../bec9c738833c7fb180256b84002dec5f/$FILE/Mobile%20Phone%20Report.pdf

Tijerina, L., Johnston, S., Parmer, E., Pham, H. A., Winterbottom, M. D., and Barickman, F. S. (2000). *Preliminary studies in haptic displays for rear-end collision avoidance system and adaptive cruise control system applications* (Report No. DOT HS 808 (TBD)). Washington, DC: National Highway Transportation Safety Administration.

Todd, J. W. (1912). *Reaction to multiple stimuli*. New York: Science Press.

Tonunaga, R. A., Hagiwara, T., Kagaya, S., and Onodera, Y. (2000). Cellular telephone conversation while driving. *Transportation Research Record, 1724,* 1–6.

Törnros, J. E. B., and Bolling, A. K. (2005). Mobile phone use–Effects of handheld and handsfree phones on driving performance. *Accident Analysis and Prevention, 37,* 902–909.

Törnros, J., and Bolling, A. (2006). Mobile phone use–Effects of conversation on mental workload and driving speed in rural and urban environments. *Transportation Research: Part F, 9,* 298–306.

Traylor, R., and Tan, H. Z. (2002). Development of a wearable haptic display for situation awareness in altered-gravity environment: Some initial findings. *Proceedings of the 10th International Symposium on Haptic Interfaces for Virtual Environment and Teleoperator Systems,* 159–164.

Treat, J. R., Tumbas, N. S., McDonald, S. T., Shinar, D., Hume, R. D., Mayer, R. E., et al. (1977). *Tri-level study of the causes of traffic accidents. Vol. 1: Causal factor tabulations and assessments. Vol. 2: Special analyses. Final Report on U.S. Department of Transportation* (Contract No. DOT-HS-034-3-535-770). Washington, DC: U.S. Government Printing Office.

Treffner, P. J., and Barrett, R. (2004). Hands-free mobile phone speech while driving degrades coordination and control. *Transportation Research Part F, 7,* 229–246.

Triggs, T. J., Lewison, W. H., and Sanneman, R. (1974). Some experiments with flight-related electrocutaneous and vibrotactile displays. In F. A. Geldard (Ed.), *Cutaneous communication systems as devices* (pp. 57–64). Austin, TX: Psychonomic Society.

Tsimhoni, O., Green, P., and Lai, J. (2001). Listening to natural and synthesized speech while driving: Effects on user performance. *International Journal of Speech Technology, 4,* 155–169.

Tufano, D. R. (1997). Automative HUDs: The overlooked safety issues. *Human Factors, 39,* 303–311.

Turatto, M., Benso, F., Galfano, G., Gamberini, L., and Umiltà, C. (2002). Non-spatial attentional shifts between audition and vision. *Journal of Experimental Psychology: Human Perception and Performance, 28,* 628–639.

Turatto, M., and Galfano, G. (2001). Attentional capture by color without any relevant attentional set. *Perception and Psychophysics, 63,* 286–297.

Turatto, M., Galfano, G., Bridgeman, B., and Umiltà, C. (2004). Space-independent modality-driven attentional capture in auditory, tactile and visual systems. *Experimental Brain Research, 155,* 301–310.

Ury, H. K., Perkins, N. M., and Goldsmith, J. R. (1972). Motor vehicle accidents and vehicular pollution in Los Angeles. *Archives of Environmental Health, 25,* 314–322.

US Senate Special Committee on Aging (1985–1986): Aging America, Trends and Projections. 1985–1986 Edn. (pp. 8–28). US Senate Special Committee on Aging (in association with the American Association of Retired Persons, the Federal Council on the Aging, and the Administration on Aging).

Van den Bos, M., Edworthy, J., Hellier, E., & Johnson, A. (2005). The effect of phonetic features on the perceived urgency of warning words in different languages. In D. De Waard, K. A. Brookhuis, R. Van Egmond, & Th. Boersema (Eds.), *Human factors in design, safety, and management* (pp. 205–218). Maastrict, The Netherlands: Shaker Publishing

van der Burg, E., Olivers, C. N. L., Bronkhorst, A. W., and Theeuwes, J. (forthcoming). Non-spatial auditory signals improve spatial visual search. *Journal of Experimental Psychology: Human Perception and Performance.*

van der Lubbe, R. H. J., and Postma, A. (2005). Interruption from irrelevant auditory and visual onsets even when attention is in a focused state. *Experimental Brain Research, 164*, 464–471.

van Erp, J. B. F. (2005). Presenting directions with a vibrotactile torso display. *Ergonomics, 48*, 302–313.

van Erp, J. B. F., Jansen, C., Dobbins, T., and van Veen, H. A. H. C. (2004). Vibrotactile waypoint navigation at sea and in the air: Two case studies. *Proceedings of EuroHaptics 2004*, 166–173.

van Erp, J. B. F., and van Veen, H. A. H. C. (2001). Vibro-tactile information processing in automobiles. *Proceedings of EuroHaptics 2001*, 99–104.

van Erp, J. B. F., and van Veen, H. A. H. C. (2003). A multi-purpose tactile vest for astronauts in the international space station. *Proceedings of EuroHaptics 2003*, 405–408.

van Erp, J. B. F., and van Veen, H. A. H. C. (2004). Vibrotactile in-vehicle navigation system. *Transportation Research Part F: Traffic Psychology and Behaviour, 7*, 247–256.

van Erp, J. B. F., and van Veen, H. A. H. C. (2006). Touch down: The effect of artificial touch cues on orientation in microgravity. *Neuroscience Letters, 404*, 78–82.

van Veen, H.-J., Spape, M., and van Erp, J. (2004). Waypoint navigation on land: Different ways of coding distance to the next waypoint. *Proceedings of EuroHaptics 2004*, 160–165.

van Veen, H. A. H. C., and van Erp, J. B. F. (2001). Tactile information presentation in the cockpit. In S. Brewster and R. Murray-Smith (Eds.), *Haptic human-computer interaction, LNCS 2058* (pp. 174–181). Berlin: Springer Verlag.

van Zomeren, A. H., and Deelman, B. G. (1976). Differential effects of simple and choice reaction after closed head injury. *Clinical Neurology and Neurosurgery, 79*, 81–90.

VCU. (2003, March 31). Traffic and fatigue, not cell phones, are top driving distractions. *Highway and Vehicle Safety Report, 3.*

Velichkovsky, B. M., Dornhoefer, S. M., Kopf, M., Helmert, J., and Joos, M. (2002). Change detection and occlusion modes in road-traffic scenarios. *Transportation Research Part F: Traffic Psychology and Behaviour, 5*, 99–109.

Verwey, W. B. (2000). On-line driver workload estimation: Effects of road situation and age on secondary task measures. *Ergonomics, 43*, 187–209.

Vinson, D. C., Mabe, N., Leonard, L. L., Alexander, J., Becker, J., Boyer, J., and Moll, J. (1995). Alcohol and injury: A case-crossover study. *Archives of Family Medicine, 4*, 505–511.

Violanti, J. M. (1997). Cellular phones and traffic accidents. *Public Health, 111*, 423–428.

Violanti, J. M. (1998). Cellular phones and fatal traffic collisions. *Accident Analysis and Prevention, 30*, 519–524.

Violanti, J. M., and Marshall, J. R. (1996). Cellular phones and traffic accidents: An epidemiological approach. *Accident Analysis and Prevention, 28*, 265–270.

Vitense, H. S., Jacko, J. A., and Emery, V. K. (2003). Multimodal feedback: An assessment of performance and mental workload. *Ergonomics, 46*, 68–87.

von Békésy, G. (1963). Interaction of paired sensory stimuli and conduction in peripheral nerves. *Journal of Applied Physiology, 18*, 1276–1284.

von Fieandt, K. (1966). Perception of the self directly and as a mirror image. In K. von Fieandt (Ed.), *The world of perception* (pp. 322–335). Homewood, IL: Dorsey.

Vroomen, J., Bertelson, P., and de Gelder, B. (2001). Directing spatial attention towards the illusory location of a ventriloquized sound. *Acta Psychologica, 108*, 21–33.

Walker, A. (1979). Music and the unconscious. *British Medical Journal, 2*, 1641–1643.

Wallace, J. S., and Fisher, D. L. (1998). Sound localization information theory analysis. *Human Factors, 40*, 50–68.

Wang, J.-S., Knipling, R. R., and Goodman, M. J. (1996). The role of driver inattention in crashes: New statistics from the 1995 crashworthiness data system. *40th Annual Proceedings of the Association for the Advancement of Automotive Medicine*, 377–392.

Ward, L. M., McDonald, J. J., and Golestani, N. (1998). Cross-modal control of attention shifts. In R. D. Wright (Ed.), *Visual attention* (pp. 232–268). New York: Oxford University Press.

Warm, J. S., Dember, W. N., and Parasuraman, R. (1991). Effects of olfactory stimulation on performance and stress in a visual sustained attention task. *Journal of the Society of Cosmetic Chemists, 42*, 199–210.

Waugh, J. D., Glumm, M. M., Kilduff, P. W., Tauson, R. A., Smyth, C. C., and Pillalamarri, R. S. (2000). Cognitive workload while driving and talking on a cellular phone or to a passenger. *Proceedings of the IEA 2000/HFES 2000 Congress*, 6-276-6-279.

Weinstein, S. (1968). Intensive and extensive aspects of tactile sensitivity as a function of body part, sex, and laterality. In D. R. Kenshalo (Ed.), *The skin senses* (pp. 195–222). Springfield, Ill.: Thomas.

Weiss, P. H., Marshall, J. C., Wunderlich, G., Tellmann, L., Halligan, P. W., Freund, H.-J., et al. (2000). Neural consequences of acting in near versus far space: A physiological basis for clinical dissociations *Brain, 123*, 2531–2541.

Wetherell, A. (1981). The efficacy of some auditory-vocal subsidiary tasks as measures of the mental load on male and female drivers. *Ergonomics, 24*, 197–214.

White, B. W. (1970). Perceptual findings with the vision-substitution system. *IEEE Transactions on Man-Machine Systems, MMS-11*, 54–58.

White, M. P., Eiser, J. R., and Harris, P. R. (2004). Risk perceptions of mobile phone use while driving. *Risk Analysis, 24*, 323–334.

Wickens, C. D. (1980). The structure of attentional resources. In R. Nickerson (Ed.), *Attention and performance VIII* (pp. 239–257). Hillsdale, NJ: Erlbaum.

Wickens, C. D. (1984). Processing resources in attention. In R. Parasuraman and D. R. Davies (Eds.), *Varieties of attention* (pp. 63–102). London: Academic Press.

Wickens, C. D. (1992). *Engineering psychology and human performance* (2nd ed.). New York: HarperCollins.

Wickens, C. D. (1999). Letter to the Editor. *Transportation Human Factors, 1*, 205–206.

Wickens, C. D. (2002). Multiple resources and performance prediction. *Theoretical Issues in Ergonomics Science, 3*, 159–177.

Wickens, C. D., and Liu, Y. (1988). Codes and modalities in multiple resources: A success and a qualification. *Human Factors, 30*, 599–616.

Wiesenthal, D. L., Hennessey, D. A., and Totten, B. (2000). The influence of music on driver stress. *Journal of Applied Social Psychology, 30*, 1709–1719.

Wiesenthal, D. L., Hennessy, D. A., and Totten, B. (2003). The influence of music on mild driver aggression. *Transportation Research Part F: Traffic Psychology and Behaviour, 6*, 125–134.

Wikman, A., Nieminen, T., and Summala, H. (1998). Driving experience and time-sharing during in-car tasks on roads of different width. *Ergonomics, 41*, 358–372.

Wilde, G. J. S. (1982). The theory of risk homeostasis: Implications for safety and health. *Risk Analysis, 2*, 209–225.

Wilkins, P. A., and Acton, W. I. (1982). Noise and accidents – A review. *Annals of Occupational Hygiene, 25*, 249–260.

Williams, A. F., Ferguson, S. A., and Wells, J. K. (2005). Sixteen-year-old drivers in fatal crashes, United States, 2003. *Traffic Injury Prevention, 6*, 202–206.

Wilson, J., Fang, M., and Wiggins, S. (2003). Collision and violation involvement of drivers who use cellular telephones. *Traffic Injury and Prevention, 4,* 45–52.

Wolfe, J. M. (1999). Inattentional amnesia. In V. Coltheart (Ed.), *Fleeting memories* (pp. 71–94). Cambridge, MA: MIT Press.

Wood, D. (1998). Tactile displays: Present and future. *Displays, 18*, 125–128.

Wood, R. L. (1988). Attention disorders in brain injury rehabilitation. *Journal of Learning Disabilities, 21,* 327–332.

Wortham, S. (1997). Are cell phones dangerous? *Traffic Safety, 97*, 14–17.

Xu, X., Wickens, C. D., and Rantanen, E. M. (2007). Effects of conflict alerting system reliability and task difficulty on pilots' conflict detection with cockpit display of traffic information. *Ergonomics, 50*, 112–130.

Yamashita, Y., Ishizawa, M., Shigeno, H., and Okada, K. (2005). The effect of adding smell into a mixed reality space: An experimental study. *Proceedings of HCI International 2005, 9*, Paper No. 1188.

Yerkes, R. M., and Dodson, J. D. (1908). The relation of strength of stimulus to rapidity of habit-formation. *Journal of Comparative Neurology and Psychology, 18*, 459–482.

Yi, D. J., Woodman, G. F., Widders, D., Marois, R., and Chun, M. M. (2004). Neural fate of ignored stimuli: Dissociable effects of perceptual and working memory load. *Nature Neuroscience, 7*, 992–996.

Yomiuri Shimbun. (2008, Feburary 25). *Computerized car-safety system test planned.* Retrieved February 26, 2008, from http://www.yomiuri.co.jp/dy/national/20080225TDY03102.htm

Young, M. S., and Stanton, N. A. (2004). Taking the load off: Investigations of how adaptive cruise control affects mental workload. *Ergonomics, 47*, 1014–1035.

Young, R. A. (2001). Association between embedded cellular phone calls and vehicle crashes involving airbag deployment. *Proceedings of the 1ˢᵗ International Driving Symposium on Human Factors in Driver Assessment, Training, and Vehicle Design*, 390–400.

Zampini, M., Brown, T., Shore, D. I., Maravita, A., Röder, B., and Spence, C. (2005). Audiotactile temporal order judgments. *Acta Psychologica, 118*, 277–291.

Zampini, M., Torresan, D., Spence, C., and Murray, M. M. (2007). Auditory-somatosensory multisensory interactions in front and rear space. *Neuropsychologia, 45*, 1869–1877.

Zeigler, B. L., Graham, F. K., and Hackley, S. A. (2001). Cross-modal warning effects on reflexive and voluntary reactions. *Psychophysiology, 38*, 903–911.

Zlotnik, M. A. (1988). Applying electro-tactile display technology to fighter aircraft – Flying with feeling again. *Proceedings of the IEEE 1988 National Aerospace and Electronics Conference*, 191–197.

Index